Brain Boosters
A Recipe for Mental Math

Walkerdoo Stimulating Math Puzzles, Volume 1

+		=	x	=	-	=
1	2	3	5	15	7	
4		8	4	32	16	
3	6	9		27	20	
	8		2	20		15

W. L. Walker

ISBN 978-1-971940-18-2 (Paperback)
ISBN 978-1-971940-19-9 (Ebook)

Inquiries and Book Orders should be addressed to:

Leavitt Peak Press
17901 Pioneer Blvd Ste L #298, Artesia, California 90701
Phone #: 2092191548

In loving memory of my darling mother, Mrs. Mattie Walker-Melton Alias "Baby Dolls"

Also, I would like to thank my wife Anna for her moral support along with staff members of Project Motivational Math, Inc.

Aim

"To show that children are born with an innate ability that can be tapped by an individual who is willing to help mold and shape their character." Our purpose is that *"Every Child Can Learn"*.

Mission Statement

To inspire each reader to reach their full potential by crunching numbers. Once achieved, they will be able to think faster, understand critical thinking, and problem solving a lot better.

**Editor: Contessa T. Walker-Jackson,
Ed.D, PMC, M.Ed., M.Ed.
Illustrator: Krystal Walker**

Table of Contents

Testimonials

Getting Students Excited about Math

<u>Grade A+ Approval</u>
"It isn't easy to get kids to learn math. But Willie Walker may have found a way."
Mary Lea Hardesty, Miami Herald

"The progressive teaching techniques and unique style has proven to be
very successful and the new standard of teaching math in Florida."
Carrie P. Meek, State Senator, District 36

"Students competing in a mental math bee acted like they were
playing a game of Nintendo rather than crunching numbers."
Sabrina Walters, Miami Herald

"As a teacher for over 20 years, and someone who enjoys teaching math,
Mr. Willie Walker left me in awe of his outstanding math ability."
Barbara Morris, Teacher, Florida City Elementary/Miami, FL.

"The excitement in the children's eyes and actions, were the highlights of
this program. Our children felt good about math and themselves."
Marilyn Jackson-Rahming, Principal,
Pineview Elementary /Tallahassee, FL

"I appreciate your efforts in assisting Dade County students to think critically in math."
Frank T. Brogan, Lt. Governor of Florida

"This letter is to inform you of a "New Look" I am seeing with the
students at Bond Elementary. This look is one of confidence in knowing,
"I can do it." Now that they have worked with you and your staff,
I can see the difference in their attitude and their work."
Barbara James, Principal
Bond Elementary / Tallahassee, Florida

"Greensboro Elementary was rated an "A" school from the state. We certainly
have to give Mr. Walker's program credit for helping us to be an "A" school."
Annette Harris, Principal,
Greensboro Elementary / Tallahassee, Florida

"After attending one of Mr. Walker's workshops in Charlotte, North Carolina I tried his methods on my students and my under-achievers out shined my over-achievers. I'm happy to say that this gives them hope"
*Sonja Eberhart, Principal, Eddleman Grade School/*Spartanburg, South Carolina

"I have used Mr. Walker's Motivational Math program and have received excellent results not only with my students in mental math computation, but the effects have carried over and have helped significantly improve their gains on standardized tests in Florida."
Stephen Herndon, B.S., M.S., Ed. S., Ph.D./
Joe Ella Goode Elementary/Miami, Florida

"Not only does Mr. Walker "aka" Mr. Math motivates children about math learning, but he is also instrumental in the development of the whole child. We enjoy Mr. Math and look forward to his next visit."
Joseph Myles, Principal/
J.D. Davis Elementary School/
Columbus, Georgia

Phase I Mental Math Pages 2-52

$$7 \times 7 + 1 + 50 - 25 =$$

$$8 \times 8 + 36 + 60 =$$

PREFACE

Welcome to Willie Walker's Wonderful World of Mental Math! Let's get started. *Nine times nine, plus 19, times four, divided by 25, take a square root, add six, multiply by the number of ounces in a pound, minus 60, plus 44, take a square root!* **The answer: 12.** How did you do? How long did it take you to answer this question? Were you able to solve the whole problem mentally without pencil or paper? Did a level of excitement take over as you visualized the problem in your head or did a feeling of fear and anxiety cause your brain to lock?

Never fear! This book will teach you the steps to performing intense number crunches like the one above. Willie Walker 'aka' Mr. Math will show you his very own style of mental math called Walker's Version. He developed this method in the classroom to focus the minds of children with very short attention spans. Mr. Math has used his version of mental math to motivate students for over 25 years. Mr. Math has years of experience preparing students for one of the hardest standardized tests in the nation called the FCAT.

Whether you are a student, teacher or parent, this book is sure to motivate you in the area of mathematics. Not only that, but in this **New Brain Power Edition** you will find lots of new information on how this program helps you to utilize the most important areas of your brain; while assisting you in the development of critical thinking skills, working memory, and number sense. Not to mention the use of short cuts and techniques that will give you speed during competitions and tests.

This book has a section that contains instructions on how to play games and do the mental exercises used in our fast-paced competitions. These exercises also called "mental gymnastics "contain a whole lot of movement, rhythm, music, visualizing and social interaction. Try these exercises in the classroom or with your family.

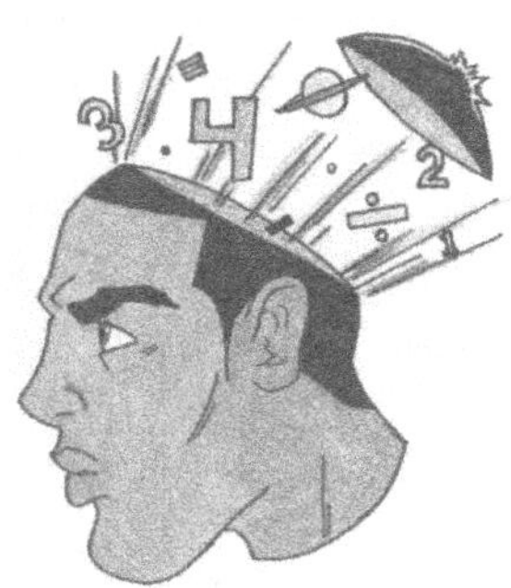

Phase I and Phase II Content

A Recipe for Mental Math is designed to motivate students to become efficient in critical thinking. This manual provides an in-depth presentation of pre-requisite skills, concepts of numbers, and problem-solving processes needed to help students become comfortable with high order thinking and be successful in basic math.

Once you have completed Phase I of this manual (Mental Math Pages (2 - 52), you should be able to:

1. Verbally recite multiplication facts from 2 - 12.
2. Add, subtract, multiply, and divide combined problems in your head.
3. Step by step method of Walker's version of mental math.
4. Use and compare counting numbers and whole numbers.
5. Use short cut methods with adding without carrying, splitting the addend, estimating, multiplying by 1/2's, 5's, 9,'s, 11's, 12's, dividing by 1/2's, 4's, and 5's, squaring numbers that end in 5's and multiplying by splitting.
6. Solve simple word problems using all four basic operations.
7. Find the square root and cube root of whole numbers.
8. Use these motivational math learning techniques to help boost brain activity.

Phase II of this manual (Walkerdoo Stimulating Math Puzzles Review Pages 53 - 164), could help with warding off dementia, even though it hasn't been proven scientifically:

Dementia presents a significant challenge as one of the fastest-growing health conditions worldwide, impacting approximately 50 million individuals with memory and cognitive issues. Brooker et al. (2019) analyzed the International Journal of Geriatric Psychiatry, which shows that puzzles like Sudoku and crosswords can help older adults stay sharp. Researchers looked at data from about 19,100 people aged 50 and older and found that those who regularly did these puzzles did better on attention, memory, and reasoning tests.

Introduction to Mental Math

In order to do Walker's version of mental math you must understand how to perform the four basic operations (addition, subtraction, multiplication, division).

Most mental math instructors will give you a one-step operation problem such as 278 plus 579, to answer in so many seconds. Walker's method starts out slowly with 3 or 4 step problems, then lead to many more. **Don't concern yourself with order of operation with Walker's version of mental math.** Whichever operation comes first is what you compute.

For example: $3 + 8 \times 5 - 5 \div 2 = 25$; first, add the 3 plus $8 = 11$; then, multiply the result by $5 = 55$; then subtract 5 from $55 = 50$; finally, divide the result by $2 = 25$.

Once you understand the pattern behind Walker's version of mental math, you will be taken to another level that leads to many repeated steps. As a mental math brain genius, you'll be able to impress your teachers, friends, family and others; and you will work faster and with more confidence on math quizzes and tests. People will be impressed with your ability to come up with the answer sometimes faster than you could have with a calculator. So, join the millions of people that are performing math the fun way with Willie Walker's version of mental math.

Sample Problems

1.   4 times 8, plus 8, divided by 5, minus 6 =   2
2.   25 minus 15, times 10, divided by 5, plus 8 =   28
3.   The square root of 169 times 5, plus 35, cut that in half, minus 8 =   42

Step by Step Method of Mental Math

Pre Test for Mental Math

Don't concern yourself with order of operation. Whichever operation comes first is what you compute.

Answers on page 44

1. $9 \times 9 + 19 = $ ______

2. $7 \times 7 + 1 + 50 + 200 \div 25 = $ ______

3. $8 \times 3 - 4 + 80 - 6 \div 10 = $ ______

4. $5 \times 5 \times 5 - 25 + 100 + 100 \div 25 \times 12 = $ ______

5. $12 \times 12 - 44 + 200 \div$ the number of seconds in a minute = ______

6. ½ of a dozen, times 6, plus 4, divided by 10 = ______

7. The number of eggs in 5 dozen, plus 40, plus 200, divided by 25, plus 8 = ______

8. ⅓ of 27, times 9, plus 19, plus 50, minus 25, times 2 = ______

9. The legs on a rectangle are 6 and 9, find the area, plus 46, plus 100, minus 4, take a square root and cut it in half = ______

10. The number of pints in a gallon, times 7, plus 44, minus 10, times 2 = ______

11. Two apples cost 35 cents; how much would 6 apples cost? Plus .95, divided by .25, times 8 = \$______

12. Find the value of y: $2y + 6 = 24$; $y = $ ______

13. The number of ounces in 2 pounds, plus 8, minus 10, times 5 divided by 15 = ______

14. The legs on a rectangular solid are 2, 4 and 6; find the volume, plus 2, times 5, plus 6, take a square root and cut it in half = ______

15. $23 \times 11 - 53 + 200 = $ ______

16. $35 \times 35 - 225 \div 25 = $ ______

17. $84 \times 5 - 20 \div 25 + 4 = $ ______

18. $54 \times 11 + 6 \div 25 + 1 \times 5 = $ ______

19. $62 \times 11 + 18 \div 25 + 2 - 5 \times 5 \times 5 = $ ______

20. $9 \times 15 + 10 + 5 \times 2 - 50 \div 25 + 8 =$ ______

21. $13 \times 13 - 69 + 400 \div 10 \times 5 + 6$, take a square root = ______

22. The cube root of 216, times 6, plus 4, times 5, plus 100, divided by the number of minutes in one hour, times 8, cut the result in half = ______

23. Add mentally: ¼, ½, 5, ¾, ½, 3¼ = ______

24. Add Mentally: $1.29, $7.41, $2.30, $5.69 = ______

25. Take the number of pounds in a ton, cut it in half and divide the result by twenty-five ______

Learn Your Multiplication Facts

Multiplication reciprocal method – Learn all in four days

Memorize 22 math facts plus perfect squares

DAY ONE

3×4 or $4 \times 3 = 12$ 4×6 or $6 \times 4 = 24$

3×6 or $6 \times 3 = 18$ 4×7 or $7 \times 4 = 28$

3×7 or $7 \times 3 = 21$ 4×8 or $8 \times 4 = 32$

3×8 or $8 \times 3 = 24$ 4×9 or $9 \times 4 = 36$

3×9 or $9 \times 3 = 27$ 4×12 or $12 \times 4 = 48$

DAY TWO

6×7 or $7 \times 6 = 42$ 7×8 or $8 \times 7 = 56$

6×8 or $8 \times 6 = 48$ 7×9 or $9 \times 7 = 63$

6×9 or $9 \times 6 = 54$ 7×12 or $12 \times 7 = 84$

6×12 or $12 \times 6 = 72$

DAY THREE

8×9 or $9 \times 8 = 72$ 9×12 or $12 \times 9 = 108$

8×12 or $12 \times 8 = 96$ 10×11 or $11 \times 10 = 110$

11×12 or $12 \times 11 = 132$

Perfect Squares

$7 \times 7 = 49$ $8 \times 8 = 64$

$2 \times 2 = 4$ $9 \times 9 = 81$

$3 \times 3 = 9$ $10 \times 10 = 100$

$4 \times 4 = 16$ $11 \times 11 = 121$

$5 \times 5 = 25$ $12 \times 12 = 144$

$6 \times 6 = 36$

$7 \times 7 = 49$

DAY FOUR – (You know already)

Zero times any number is 0.   Example: $4 \times 0 = 0$

One times any number is that number: Example: $9 \times 1 = 9$

Two's are easy. Just keep adding two.   Example: 2, 4, 6, 8, 10, 12….

Five's are easy. Just keep adding five.   Example: 5, 10, 15, 20, 25, 30….

Tens are easy. Just keep adding ten.   Example: 10, 20, 30, 40, 50….

Eleven's are easy. Just write the same number twice.   Example: $3 \times 11 = 33$; $5 \times 11 = 55$; $9 \times 11 = 99$

Splitting and Rounding the Addend

Some numbers are hard to add without splitting the number
into parts or rounding to powers of 10's.

Adding and Subtracting using 5's and 10's

1.	$1 + 4 = 5$	6.	$2 + 8 = 10$	11.	$10 - 2 = 8$
2.	$2 + 3 = 5$	7.	$3 + 7 = 10$	12.	$10 - 3 = 7$
3.	$5 - 3 = 2$	8.	$4 + 6 = 10$	13.	$10 - 4 = 6$
4.	$5 - 4 = 1$	9.	$5 + 5 = 10$	14.	$10 - 5 = 5$
5.	$1 + 9 = 10$	10.	$10 - 1 = 9$		

Adding numbers to powers of 10's:

1.	$5 + 5 = 10$	7.	$7 + 13 = 20$	13.	$9 + 11 = 20$
2.	$15 + 5 = 20$	8.	$17 + 13 = 30$	14.	$19 + 11 = 30$
3.	$25 + 5 = 30$	9.	$27 + 13 = 40$	15.	$29 + 11 = 40$
4.	$6 + 4 = 10$	10.	$8 + 12 = 20$	16.	$10 + 10 = 20$
5.	$16 + 4 = 20$	11.	$18 + 12 = 30$	17.	$20 + 10 = 30$
6.	$26 + 4 = 30$	12.	$28 + 12 = 40$	18.	$30 + 10 = 40$

Splitting the addend by adding and subtracting:

1.	$15 + 6 = ?$	$15 + 5 + 1 = 21$	11.	$68 + 19 = 68 + 20 - 1 = 87$
2.	$18 + 12 = ?$	$18 + 10 + 2 = 30$	12.	$35 + 27 = 35 + 30 - 3 = 62$
3.	$25 + 6 = ?$	$25 + 5 + 1 = 31$	13.	$57 + 18 = 57 + 20 - 2 = 75$
4.	$37 + 13 = ?$	$37 + 10 + 3 = 50$	14.	$82 + 13 = 82 + 10 + 3 = 95$
5.	$28 + 7 = ?$	$28 + 2 + 5 = 35$	15.	$26 + 16 = 26 + 10 + 6 = 42$
6.	$39 + 9 = ?$	$40 + 10 - 2 = 48$	16.	$79 + 43 = 80 + 40 - 1 + 3 = 122$
7.	$54 + 14 = ?$	$54 + 10 + 4 = 68$	17.	$64 + 39 = 60 + 40 + 4 - 1 = 103$
8.	$57 + 18 = ?$	$57 + 20 - 2 = 75$	18.	$97 + 70 = 100 + 70 - 3 = 167$
9.	$19 + 12 = ?$	$19 + 10 + 2 = 31$	19.	$73 + 56 = 70 + 60 + 3 - 4 = 129$
10.	$36 + 25 = ?$	$36 + 20 + 5 = 61$	20.	$59 + 62 = 60 + 60 - 1 + 2 = 121$

Your Time!

Sample: 8 + 12 = ?   8 + 10 + 2 = 20 or 8 + 2 + 10 = 20

1. 7 + 11 =	11. 14 + 15 =
2. 9 + 6 =	12. 28 + 16 =
3. 23 + 8 =	13. 35 + 15 =
4. 15 + 8 =	14. 37 + 11 =
5. 17 + 13 =	15. 42 + 8 =
6. 26 + 9 =	16. 44 + 16 =
7. 35 + 14 =	17. 48 + 19 =
8. 54 + 16 =	18. 55 + 14 =
9. 49 + 12 =	19. 56 + 12 =
10. 58 + 16 =	20. 63 + 37 =

Mental Math Concepts With Money

Everyone can count money, even if they do not understand many math concepts. If you can multiply you can divide, if you can add you can subtract. Ordinarily, when you multiply you can basically compute all four operations. Let's see how multiplication and division problems work together! How many quarters does it take to make a dollar? The answer is 4. How many quarters does it take to make two dollars? The answer is *8*, and so on. Can you find the pattern? If it takes 4 quarters to make a dollar, then $100 \div 25$ is 4, or $200 \div 25$ is 8. When you are dividing hundreds by 25, just multiply the first digit by 4. For example: $\underline{1}00 \div 25 = 4$ or $(\underline{1} \times 4 = 4)$, $\underline{2}00 \div 25 = 8$ or $(\underline{2} \times 4 = 8)$, and $\underline{3}00 \div 25 = 12$ or $(\underline{3} \times 4 = 12)$.

It takes 4 quarters to make $1.00

25¢
25¢
25¢
25¢
____
$1.00

How many quarters does it take to make? *Think!*

$2 = ____________ $2.75 = ____________

$3 = ____________ $3.50 = ____________

$4 = ____________ $4.25 = ____________

$5 = ____________ $5.75 = ____________

$6 = ____________ $15 = ____________

$7= ____________ $16 = ____________

$8= ____________ $18 = ____________

$9= ____________ $19 = ____________

$10= ____________ $20 = ____________

Mental Math Problems

Forget about order of operation: Whichever operation
comes up first is what you perform!

Three Step Mental Math Problems

Example A: 2 + 2 − 1 x 3

To Solve:

Step 1   2 + 2 = 4 (Add)

Step 2   4 - 1= 3 (Subtract)

Step 3   3 x 3 = 9 (Multiply)

The answer is <u>9</u>

****

Put on your thinking caps. You try these!

Solve the problems:

1)   4 + 2 − 3 x 2=   7)   9 x 9 + 19 − 6=

2)   6 x 2 − 2 + 8=   8)   12 x 12 − 44 + 87=

3)   7 + 4 − 1 x 3=   9)   55 − 15 x 2 + 10=

4)   6 − 2 x 4 + 4=   10)   12 ÷ 2 + 4 x 6=

5)   7 x 5 + 5 − 2=   11)   75 − 25 + 8 + 2=

6)   10 − 3 x 7 + 1=   12)   25 + 30 − 5 x 2=

Good Job! ☺

Three Step Mental Math Problems

Example B: $4 \times 4 + 4 - 5 =$

To Solve:

Step 1	$4 \times 4 = 16$ (Multiply)
Step 2	$16 + 4 = 20$ (Add)
Step 3	$20 - 5 = 15$ (Subtract)

The answer is <u>15</u>

****

Here we go! Try it!

Solve the problems:

1) $4 + 4 + 4 \times 3 =$

2) $6 - 1 \times 5 + 7 =$

3) $10 \times 10 - 5 + 6 =$

4) $8 \times 7 + 4 - 5 =$

5) $4 \times 3 + 3 + 5 =$

6) $18 - 8 \times 5 - 5 =$

7) $8 \times 8 + 36 - 8 =$

8) $10 - 6 + 12 \div 2 =$

9) $14 + 6 + 5 \div 5 =$

10) $9 \times 9 + 9 + 10 =$

11) $11 \times 11 - 21 + 63 =$

12) $12 \times 12 + 6 - 25 =$

Great! ☺

Three Step Mental Math Problems

Example C: $12 - 6 \times 3 + 2 =$

To Solve:

Step 1	$12 - 6 = 6$ (Subtract)
Step 2	$6 \times 3 = 18$ (Multiply)
Step 3	$18 + 2 = 20$ (Add)

The answer is <u>20</u>

****

Your time!

Solve the problems:

1) $5 \times 5 - 5 + 15 =$

2) $8 - 3 \times 5 + 6 =$

3) $14 - 4 \times 10 + 21 =$

4) $6 + 4 \times 5 + 8 =$

5) $7 \times 3 + 0 + 6 =$

6) $8 \times 5 + 6 + 4 = 12)$

7) $4 \times 4 + 10 - 1 =$

8) $12 \times 8 + 28 - 16 =$

9) $10 \times 10 - 25 + 10 =$

10) $24 \div 6 + 6 \times 8 =$

11) $30 \div 5 + 4 \times 9 =$

$17 - 10 \times 6 + 8 =$

Wow! ☺

Four Step Mental Math Problems

Example A: $5 \times 5 - 6 + 11 - 6 =$

To Solve:

Step 1	**5 x 5 = 25 (Multiply)**
Step 2	**25 – 6 = 19 (Subtract)**
Step 3	**19 + 11 = 30 (Add)**
Step 4	**30 – 6 = 24 (Subtract)**
	The answer is <u>24</u>

****

Go for it!

Solve the problems:

1) $7 \times 7 + 1 - 10 + 8 =$

2) $3 \times 3 + 11 - 5 + 10 =$

3) $7 + 4 + 4 - 5 \times 10 =$

4) $12 - 8 \times 4 + 4 + 8 =$

5) $17 - 2 \times 4 - 10 + 7 =$

6) $8 \times 4 - 2 + 6 + 4 =$

7) $15 + 5 - 10 \times 10 + 37 =$

8) $9 \times 7 + 37 - 25 + 8 =$

9) $27 - 7 \times 5 - 100 + 16 =$

10) $14 + 6 \div 2 \times 10 + 9 =$

11) $1/2 \text{ of } 10 \times 5 + 8 + 2 =$

12) $1/2 \text{ of } 20 - 5 \times 10 + 6 =$

Fantastic! ☺

Four Step Mental Math Problems

Example B: $8 + 18 - 6 - 5 + 15 =$

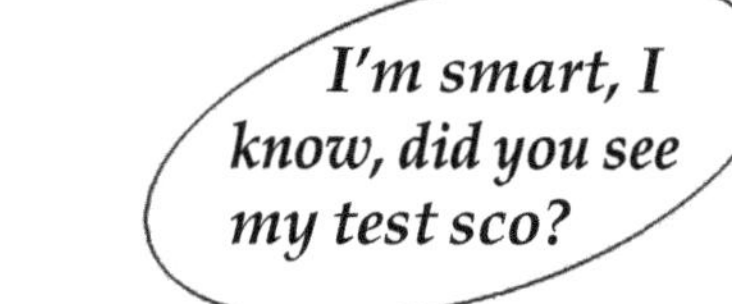

To Solve:

Step 1	$8 + 18 = 26$ (Add)
Step 2	$26 - 6 = 20$ (Subtract)
Step 3	$20 - 5 = 15$ (Subtract)
Step 4	$15 + 15 = 30$ (Add)

The answer is <u>30</u>

****

These are for you!

Solve the problems:

1) $19 + 6 - 5 + 8 + 7 =$

2) $25 - 5 + 40 + 6 - 7 =$

3) $9 \times 9 + 19 - 19 + 15 =$

4) $6 \times 8 + 2 \times 3 - 50 =$

5) $7 \times 9 + 37 - 50 + 25 =$

6) $15 + 35 - 40 \times 10 + 13 =$

7) $18 + 12 - 10 \div 2 \times 4 =$

8) $17 + 13 + 20 - 25 + 5 =$

9) $1/2$ of $24 \times 12 - 44 + 18 =$

10) $15 - 5 \times 10 - 50 + 27 =$

11) $11 \times 11 - 21 + 200 \div 25 =$

12) $7 \times 7 + 1 + 50 - 75 =$

Tenacious! ☺

Four Step Mental Math Problems

Example C: $16 - 6 \times 10 - 19 \div 9 =$

To Solve:

Step 1	$16 - 6 = 10$ (Add)
Step 2	$10 \times 10 = 100$ (Multiply)
Step 3	$100 - 19 = 81$ (Subtract)
Step 4	$81 \div 9 = 9$ (Divide)

The answer is <u>9</u>

****

Go for it!

Solve the problems:

1) $8 \times 6 + 2 - 25 - 5 =$ _____

2) $15 + 5 \times 5 - 25 - 5 =$ _____

3) $16 + 4 - 10 \times 10 + 37 =$ _____

4) $75 - 25 \times 2 + 50 - 25 =$ _____

5) $15 \times 4 + 40 + 100 \div 25 =$ _____

6) $18 + 12 \times 3 \div 10 \times 9 =$ _____

7) $6 \times 8 + 2 + 50 - 30 =$ _____

8) $7 \times 8 + 44 + 100 \div 25 =$ _____

9) $6 \times 6 + 4 + 60 + 200 =$ _____

10) $9 \times 9 - 81 + 87 + 13 =$ _____

11) $1/2$ of $50, \div 5 + 6 \times 11 =$ _____

12) $1/3$ of $30, \times 10 + 100 \div 25 =$ _____

Super!

Five Step Mental Math Problems

Example: 2/3 of 24 + 4 x 5 – 6 ÷ 10 =

To Solve:

Step 1   2/3 of 24 = 16
          (24 ÷ 3 = 8) (8 x 2 = 16)

Therefore, the answer is 16

Step 2   16 + 4 = 20 (Add)
Step 3   20 x 5 = 100 (Multiply)
Step 4   100 – 6 = 94 (Subtract)
Step 5   94 ÷ 10 = 9.4 (Divide)
          The answer is <u>9.4</u>

****

Go for it! 

Solve the problems: (Remember no pencil and paper or calculator)!

1)   9 x 6 + 46 – 25 ÷ 15 x 3 = _____   7)   14 ÷ 2 + 3 + 50 – 30 + 18 = _____

2)   15 x 5 + 5 x 2 – 5 + 23 = _____   8)   26 – 17 x 9 + 19 + 100 ÷ 25 = _____

3)   16 + 4 – 10 x 10 + 37 – 5 = _____   9)   2.4 x 10 + 6 x 6 + 20 + 60 = _____

4)   75 – 25 x 2 + 50 – 25 + 8 = _____   10)   9 x 9 – 81 + 87 + 13 + 67 = _____

5)   15 x 4 + 40 – 50 ÷ 25 - 12 = _____   11)   11/2 of 50 ÷ 5 + 6 x 11 + 19 = _____

6)   20 + 18 + 12 x 3 ÷ 10 x 6 = _____   12)   1/3 of 27 x 10 + 100 + 10 ÷ 25 = _____

Great!

Six Step Mental Math Problems

Example: $57 + 3 \times 2 \div 12 \times 10 + 200 \div 25 =$

To Solve:

Step 1 $57 + 3 = 60$ (Add)

Step 2 $60 \times 2 = 120$ (Multiply)

Step 3 $120 \div 12 = 10$ (Divide)

Step 4 $10 \times 10 = 100$ (Multiply)

Step 5 $100 + 200 = 300$ (Add)

Step 6 $300 \div 25 = 12$ (Divide)

The answer is <u>12</u>

****

You can do it!

1) $17 \times 2 + 6 + 10 \times 2 + 100 - 25 =$ ________

2) $15 \times 5 + 25 \div 25 + 96 \times 10 + 69 =$ ________

3) $8 \times 8 + 36 + 37 - 7 \div 13 \times 8 =$ ________

4) $100 - 5 + 5 + 5 + 75 + 50 - 25 =$ ________

5) $80 \div 2 \times 5 + 100 + 100 \div 25 - 10 =$ ________

6) $7 \times 8 + 44 \div 10 \times 9 \div 10 + 4 =$ ________

7) $6 \times 8 + 12 + 40 \div 10 \times 10 + 27 =$ ________

8) 10×10, 1/2 of that, $+ 70 + 30 - 25 + 6 =$ ________

Great Job! ☺

Launching Into Orbit with Step By Step Practice Problems

1. 3 x 5 = 15 minus 5 = 10

2. 6 plus 6 = 12 plus 8 = 20 minus 4 = 16

3. 5 x 5 = 25 minus 5 = 20 plus 11 = 31

4. 4 x 5 = 20 minus 5 = 15 plus 6 = 21

5. 6 x 5 = 30 minus 8 = 22 plus 6 = 28

6. 3 x 3 = 9 plus 1 = 10 x 10 = 100

7. 4 x 4 = 16 minus 6 = 10 x 5 = 50 plus 12 = 62

8. 8 minus 5 = 3 x 8 = 24 plus 1 = 25 minus 5 = 20

9. 17 minus 7 = 10 x 10 = 100 divided by 10 = 10 plus 13 = 23

10. 16 minus 6 = 10 plus 8 = 18 plus 2 = 20 divided by 4 = 5 x 5 = 25

11. 7 plus 3 = 10 minus 4 = 6 x 6 = 36 plus 4 = 40 plus 7 = 47

12. 18 plus 7 = 25 divided by 5 = 5 x 6 = 30 plus 16 = 46

13. 8 x 8 = 64 plus 6 = 70 minus 10 = 60 divided by 10 = 6

14. 30 divided by 6 = 5 x 5 = 25 minus 5 = 20 plus 11 = 31

15. 9 x 9 = 81 plus 19 = 100 divided by 10 = 10 plus 40 = 50 minus 25 = 25

16. 26 minus 6 = 20 plus 80 = 100 divided by 10 = 10 x 10 = 100 plus 37 = 137

17. 200 divided by 25 = 8 times 4 = 32 plus 18 = 50 x 6 = 300

18. 59 minus 9 = 50 x 5 = 250 plus 25 = 275 plus 25 = 300 divided by 25 = 12

19. 20 + 87 = 107 minus 7 = 100 x 4 = 400 divided by 20 = 20

20. 6 x 6 + 4 + 60 − 50 x 5 + 50 ÷ 25 + 3 x 5 + 6 =?

More Practice Problems

1. 5 times 5 plus 8 =

2. 7 plus 8 minus 5 times 4 =

3. 20 plus 10 minus 5 plus 6 =

4. 9 plus 7 minus 6 plus 12 =

5. 20 minus 10 times 4 plus 9 =

6. 7 times 7 plus 1 plus 27 =

7. 8 times 6 plus 2 minus 25 plus 6 =

8. 4 times 8 plus 8 divided by 8 times 5 =

9. 6 times 6 plus 4 plus 60 divided by 10 =

10. 8 plus 9 plus 3 divided by 4 times 5 plus 7 =

11. 9 times 7 minus 3 plus 40 divided by 25 times 8 =

12. 6 times 5 divided by 5 times 6 plus 4 plus 10 plus 25 =

13. 8 times 8 plus 6 divided by 10 times 7 plus 1 minus 25 =

14. 9 times 9 plus 19 plus 100 divided by 25 times 5 plus 6 =

15. 12 times 12 minus 44 plus 100 divided by 25 plus 3 times 5 =

16. 7 times 7 plus 1 plus 50 plus 21 take a square root minus 8 =

17. 36 minus 16 plus 5 plus 175 divided by 25 plus 9 =

18. The number of eggs in 6 dozens plus 28 minus 23 plus 15 =

19. 1/3 of 24 times 8 plus 36 plus 200 divided by 25 plus 16 =

20. 20% of 100 plus 75 divided by 10 + 10.5 =

Tables of Measure – Study and Memorize
Study and Memorize

1 dozen = 12   8 ounces = 1 cup
1 pint = 16 ounces   16 ounces = 2 cups or 1 pint
1 quart = 2 pints   32 ounces = 4 cups or 2 pints
1 gallon = 4 quarts   4 cups = 2 pints or 1 quart or ¼ gallon
1 peck = 8 quarts   8 cups = 4 pints or 2 quarts or ½ gallon
1 bushel = 4 pecks   12 cups = 6 pints or 3 quarts or ¾ gallon
1 pound = 16 ounces   16 cups = 8 pints or 4 quarts or 1 gallon

1 foot = 12 inches   1 ounce = 2 tablespoons
1 yard = 3 feet = 36 inches   2 ounces = ¼ cup
1 rod = 16. feet = 5½ yards   4 ounces = ½ cup

1 minute = 60 seconds
1 hour = 60 minutes
1 day = 24 hours
1 year = 12 months = 52 weeks = 365 ¼ days

1 mile = 5,280 feet = 1,760 yards
1 acre = 4,840 square yards = 43,560 square feet
1 ton = 2,000 pounds

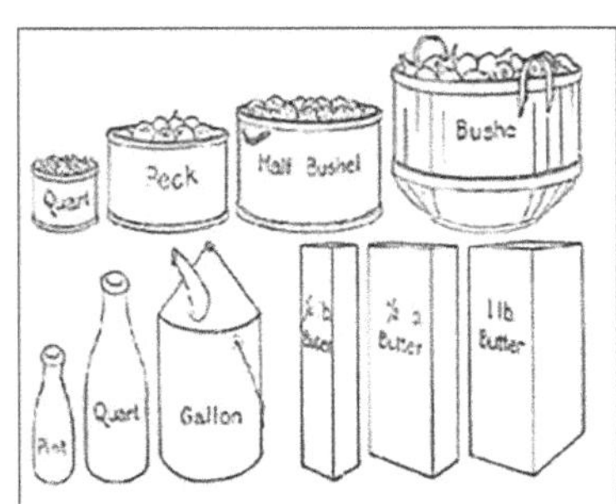

Sample Problems

1. The number of pints in a quart times 8 + 4 minus the number of quarts in a gallon = ?

    Answer is 16: 2 is the number of pints in a quart times 8 equal 16, plus 4 equal 20, minus 4 (4 is the number of quarts in a gallon) = 16.

2. The number of pecks (4) in a bushel plus the number of ounces (16) in a pound minus 5 × 4 + 40 ÷ by the number of pints in a quart =

    Answer: 50

Measurement Problems

1. 1/2 of a dozen, plus 14, times 5, minus 50, times 3, plus 10, divided by the number of ounces in a pound, plus 15 =

2. The number of pounds in a ton minus 1000, cut that in half, divided by 25, times the number of eggs in two dozen =

3. Eight times the number of quarts in 3 gallons, plus 4, cut that in half, cut that in half, plus 5, divided by 6 =

4. Nine times 9, plus nineteen, plus 100, plus 100, divided by the number of seconds in a minute, plus 5, times five, plus 100, divided by 15, plus 3 =

5. 1/4 of two dozen, times 7, plus 8, times 5, divided 25, times 12, 1/3 of that, plus 8, divided by 8 =

6. Six times 8, plus 52, plus 100, divided by 25, times 8, plus 36, plus 200, divided by the number of minutes in an hour, times 5, times 5, minus 4, take the square root, times 6, plus 2 =

7. 3/8 of two dozen, times 9, plus nineteen, plus 44, take the square root, times 5, plus 40 plus 69, plus 31, minus 49 =

8. The number of inches in 3 feet plus 14, times 6, plus 300, divided by the number of seconds in a minute, times 20, cut that in half, cut that in half, divided by 5, plus 13 =

9. 1/2 of the number of ounces in a pound times 8, plus 36, minus 36, plus 6, divided by 10, minus 2, times 6, plus12, divided by 6, plus 3, times 10 =

10. 15 times 5, plus 25, plus 69, take the square root, plus 2, times 5, minus 25, plus 6, minus 6, minus 1, divided by 7, times 8, plus 44, minus 8 =

Solve:

8 pints =_______ gallons 8 gallons = _______quarts 2 tons = _______ pds

24 inches = _______ feet 6 yards = _______feet 36 inches = _______yds

120 sec = _______ mins 240 mins = _______ hrs 48 hrs = _______ days

233 lbs 5 oz	6 hrs 28 mins 33 sec	6 yds 9 ft 8 in
− 158 lbs 12 oz	− 4 hrs 30 mins 45 sec	+ 4 yds 6 ft 7 in

Squaring and Cubing Numbers 2 - 20

Study and Memorize!!

<u>Squares</u>

1. $2^2 = 2 \times 2 = 4$

2. $3^2 = 3 \times 3 = 9$

3. $4^2 = 4 \times 4 = 16$

4. $5^2 = 5 \times 5 = 25$

5. $6^2 = 6 \times 6 = 36$

6. $7^2 = 7 \times 7 = 49$

7. $8^2 = 8 \times 8 = 64$

9. $10^2 = 10 \times 10 = 100$

10. $112 = 11 \times 11 = 121$

11. $122 = 12 \times 12 = 144$

12. $132 = 13 \times 13 = 169$

13. $14^2 = 14 \times 14 = 196$

14. $15^2 = 15 \times 15 = 225$

15. $16^2 = 16 \times 16 = 256$

16. $17^2 = 17 \times 17 = 289$

17. $18^2 = 18 \times 18 = 324$

18. $192 = 19 \times 19 = 361$

19. $202 = 20 \times 20 = 400$

<u>Cubes</u>

1. $23 = 2 \times 2 \times 2 = 8$

2. $33 = 3 \times 3 \times 3 = 27$

3. $43 = 4 \times 4 \times 4 = 64$

4. $53 = 5 \times 5 \times 5 = 125$

5. $63 = 6 \times 6 \times 6 = 216$

6. $73 = 7 \times 7 \times 7 = 343$

10. $113 = 11 \times 11 \times 11 = 1331$

11. $123 = 12 \times 12 \times 12 = 1728$

12. $133 = 13 \times 13 \times 13 = 2197$

13. $143 = 14 \times 14 \times 14 = 2744$

14. $153 = 15 \times 15 \times 15 = 3375$

15. $163 = 16 \times 16 \times 16 = 4096$

7. 83 = 8 X 8 X 8 = 512

16. 173 = 17 X 17 X 17 = 4913

8. 93 = 9 X 9 X 9 = 729

17. 183 = 18 X 18 X 18 = 5832

9. 93 = 9 X 9 X 9 = 729

19. 193 = 19 X 19 X 19 = 6859

20. 203 = 20 X 20 X 20 = 8000

Tips Before Performing Advance Mental Math
Study

<u>Square Roots</u> $\sqrt{4}$ 2 x 2 = 4 Therefore, the square root is 2

A number times itself has to equal the number under the radical sign.

Example:

a) $\sqrt{25}$ 5 x 5 = 25 Ans. (5)   b) $\sqrt{64}$ 8 x 8 = 64 Ans. (8)
c) $\sqrt{144}$ 12 x 12 = 144 Ans. (12)   d) $\sqrt{196}$ 14 x 14 = 196 Ans. (14)

<u>Cube Roots</u> $3\sqrt{}$

Cube roots are much like square roots, however, taken a few steps further.

2^3 = 2 x 2 x 2 = 8, the 2 is called the base and the 3 is called the exponent or power. Therefore, the cube root of 8 is 2.

Example:

A   The cube root of 27 is 3 or (33) is 3 x 3 x 3 = 27

B   The cube root of 64 is 4 or (43) is 4 x 4 x 4 = 64

C   The cube root of 125 is 5 or (53) is 5 x 5 x 5 = 125

<u>Exponents or Powers</u>

The number of times the base is multiplied by itself.

Example:

A   23 or 2 to the third power is 2 x 2 x 2 = 8

B   34 or 3 to the fourth power is 3 x 3 x 3 x 3 = 81

C   54 or 5 to the fourth power is 5 x 5 x 5 x 5 = 625

ADVANCE MENTAL MATH
Cubes, Squares, Exponents, etc.

$$9.7 \times 10 + 3 = ?$$

Congratulations, you've made it to the advanced mental math section!!

Advanced Mental Math Problems

1) $6 \times 6 + 4 \times 5 - 75 \div 25 \times 7 + 15 =$

2) $8 \times 4 + 18 - 10 \times 5 + 100 + 100 \div 25 + 4 \times 5 + 100 \div 25 \times 4 =$

3) $15 \times 5 + 75 + 50 + 300 \div 25 \times 5 + 75 - 25 \times 2 \div 25 + 3 \times 5 - 4 =$

4) $30 \times 5 + 150 \div 25 + 3 \times 5 + 25 - 75 \times 5 + 25 \div 15 \times 10 - 8 + 3 - 14 \div 9 =$

5) $4 \times 3 \text{ x } 12 + 6 \div 15 \times 10 + 37 + 13 + 150 \div 25 + 2 \times 2 + 2 \times 3 \div 10 \times 9 + 19 =$

6) $5 \times 6 \times 3 \div 10 \times 9 + 4 \div 10 + .5 \times 8 + 28 + 100 \div 25 \times 4 + 18 + 6 + 4 \div 10 + 8 =$

7) $15 \times 4 + 60 + 80 - 75 \div 25 \times 5 + 75 \div 10 \times 4 + 50 \div 9 \times 4 + 2 \div 7 \times 8 \div 10 =$

8) $14 \times 2 + 2 \times 4 + 30 \div 15 \times 5 \times 5 + 50 + 300 \div 25 + 6 \times 3 + 90 + 20 \div 25 \times 4 + 8$
     $\times 5 + 15 - 7 =$

9) $15 \times 6 \div 9 \times 10 + 43 + 7 \div 15 \times 10 + 56 + 4 - 60 + 32 + 18 + 25 \div 25 \times 7 + 1 \times 6$
     $+ 200 \div 25 \times 20 =$

10) $4 \times 9 + 14 \times 5 + 50 + 100 \div 25 + 4 \times 5 - 75 \times 5 + 50 \div 25 \times 7 + 1 \div 10 \times 5 + 7 +$
     $3 \times 2 + 8 \div 10 =$

11) The square root of $25 \times 5 + 75 + 85 + 15 \div 25 \times 5 \times 5 \div 10 - 5 \times 6 + 8 \div 10 =$

12) The square root of $64 \times 5 + 60 \div 10 + 57 + 3 \div 7 \times 4 + 18 \div 10 + .2 \times 4 \div 10 =$

13) The cube root of $27 \times 3 \times 9 + 19 \div$ by the cube root of $125 \times 5 + 100 - 25 \div 25 =$

14) The cube root of $64 \times 16 + 36 \div$ by the square root of $25 + 30 - 5 \times 2 + 7 \div 10 = 15$

15) $2^4 \times 4 \div 10 + .6 \times 7 + 1 \times 5 + 50 + 100 - 150 \div 25 \times 8.1 + 19 \div 25 \text{ x } 6 + 8 =$

16) $3^3 + 27 + 6 - 30 \div 6 \times 5 \times 5 \div 25 \times 15 + 30 + 45 \div 15 \times 7 + 130 \div 10 \times 5 + 80 \div$ by
     the # of seconds in a minute $\times 8 + 1 \times 5 - 26 =$

17) $15^2 + 75 \div 15 \times 5 + 100 + 200 + 100 - 250 \div 2 \times \sqrt{4} \div 10 \times 4 + 59 - 4 =$

18) $13^2 + 31 \div 25 \times 8 \div \sqrt{64} \times 5 \times 4 + 40 \div 5 \times 5 \div 20 \times 40 \div 25 =$

19) $18 \times 18 + 26 - 150 \div 25 \times 9 + 28 - 10 \times 2 + 20 \div 10 \times 10 \div 25 \times 4 + 8 - 10 + 15 \div 10 + 8 =$

20) $180 - 30 \div$ by the $\sqrt{225} + 10 \times 10 + 400 + 300 \div 2 + 50 \times 6$ minus the # of pounds in 1 ton $+ 125 =$

21) The # of seconds in 2 minutes $\times 2 + 60 \div 10 - 15 \times 5 + 5 \times 4 \div 8 + 13 + 7 \times 3 \div 20 \times 9 + 9 \times 2 + 20 - 75 =$

22) The # of inches in 3 feet $x 2 + 28 \div 10 + 37 + 3 \div 5 - 3 \times 8 + 44 + 89 - 9 + 27 - 7 \div 25 =$

23) $2 \times 17 + 6 - 15 \times 6 \div 15 \times 8 \div 8 \times 12 + 30 + 17 - 7 + 40 - 125 \times 2 + 50 - 19 - 31 \div 15 + 5 \times 6 \div 2 + 8 + 2 \times 2 =$

24) $56 - 16 \times 5 \div 25 + 12 \times 10 + 100 + 100 \div 25 + 4 - 8 \times 12 - 14 \div 13 + 27 + 3 \times 6 - 20 + 30 \div 25 + 86 + 4 - 78 + 3 \times 7 + 14 =$

25) $14 \times 14 + 4 + 400 \div 25 + 1 \times 100$ minus the # of pounds in a ton $\div 25 \times 8 + 17 - 7 \times 2 + 10 + 57 - 7 \div 20 + 40 + 65 + 8 - 8 \div 25 =$

26) $3/4$ of $100 \div 25 \times 5 \times 5 + 25$ take the square root $+ 90 + 100 \div 25 =$

27) Find the volume of a cube that has 7cm as the length, 20cm as the width and 10cm as the height minus $1000 \div 25 + 4 + 400 - 60 \div$ the # of seconds in a minute $+ 97 =$

28) 10% of $200 \times 5 + 300 - 150 \div 25$ x the # of eggs in a dozen $+ 1$ take the square root $\times 34 =$

29) $7/10$ of $100 + 30 + 44$ take the square root $x 13 + 44 - 20 \div 18 +$ the # of hours in a day $\times 2 + 12 \div 10 \times 9 - 22 =$

30) If 2 apples cost 35, then 6 apples would cost_____, $+ 95 - 50 x 2 \div 25 x 11 + 18 + 75 \div 25 \times 9 + 19 + 200 x 8 -$ the # of pounds in a ton $\div 20 \times 5 + 25$ take the cube root $\times 15 + 25 - 75 + 5 - 8 \times 2 =$

Motivational Math Short Cut Methods

Lesson 1 ADDING NUMBERS WITHOUT CARRYING

The watch maker must assemble 300 gold watches to ship to the jeweler. During the next four days, he assembled 32 watches, 26 watches, 38 watches and 44 watches. He would like to figure out the total number of watches he assembled. To solve this problem, you would add 32 + 26 + 38 + 44.

Directions: When adding numbers without carrying always start from the left column. First add the tens digit from the left column by remembering your total after each problem.

Problem   32 + 26 + 38 + 44 = ?
Tens Column

3 2 add 30 from left column (Round to the power of 10's)
Step 1:   2 6 plus 20 = 50
3 8 plus 30 = 80
+ 4 4 plus 40 = 120

Then add to your total each number from the ones digit from the right column.

Ones Column

3 2 add 120 + 2 = 122
Step 2:   2 6 add 122 + 6 = 128
3 8 add 128 + 8 = 136
+ 4 4 add 136 + 4 = 140

Answer: 140

****

Practice Problems:

1.	32	2.	43	3.	54	4.	123
	44		37		46		235
	38		63		65		684
	+ 75		+ 84		+ 78		+ 891

Lesson 2 MULTIPLYING AND DIVIDING BY 4

Multiplying

DIRECTIONS: When multiplying by 4, first double the number, and then double it again.

Example A: 16 x 4

Step 1:   Double 16 by multiplying by 2. ⟶ 16 x 2 = 32

Step 2:   Double the 32 by multiplying it by 2. ⟶ 32 x 2 = 64

Answer: 64

Practice Problems:

1. 32 x 4 =
2. 28 x 4 =
3. 72 x 4 =
4. 36 x 4 =
5. 44 x 4 =

6. 50 x 4 =
7. 62 x 4 =
8. 18 x 4 =
9. 75 x 4 =
10. 128 x 4 =

****

Dividing

Directions: When dividing by 4, first cut the number in half, then cut it in half again.

Example B: 56 ÷ 4

Step 1:   Cut 56 in half by dividing it by 2.   56 ÷ 2 = 28
Step 2:   Cut 28 in half by dividing it by 2.   28 ÷ 2 = 14
Answer: 14

Practice Problems:

1. 32 ÷ 4 =
2. 28 ÷ 4 =
3. 72 ÷ 4 =
4. 36 ÷ 4 =
5. 44 ÷ 4 =

6. 56 ÷ 4 =
7. 64 ÷ 4 =
8. 82 ÷ 4 =
9. 96 ÷ 4 =
10. 128 ÷ 4 =

Lesson 3 MULTIPLYING BY 5

Multiplying Even Numbers:

DIRECTIONS: When multiplying an **EVEN** whole number, first cut the number in half by dividing by 2, then place a zero at the end of that number.

Example A: Multiplying a whole EVEN number by 5: ⟶ 16 x 5

Step 1: Cut the 16 in half by dividing it by 2. ⟶ $16 \div 2 = 8$

Step 2: Place a zero behind the 8. ⟶ 8 (0) = 80

Answer: 80

****

*** Multiplying Odd Numbers:

DIRECTIONS: When multiplying an ODD whole number, first take the preceding whole number and cut it in half and then add a five at the end of the number.

Example B: Multiplying a whole **ODD** number by 5:   15 x 5

Step 1: Find the preceding number by subtracting one. $15 - 1 = 14$

Step 2: Cut the 14 in half by dividing by 2. $14 \div 2 = 7$

Step 3: Place a five behind 7   7 (5) = 75

Answer: 75

Practice Problems:

1. 18 x 5 =	6. 24 x 5 =	11. 15 x 5 =
2. 20 x 5 =	7. 84 x 5 =	12. 21 x 5 =
3. 28 x 5 =	8. 32 x 5 =	13. 33 x 5 =
4. 44 x 5 =	9. 56 x 5 =	14. 61 x 5 =
5. 64 x 5 =	10. 70 x 5 =	15. 83 x 5 =

Lesson 4 MULTIPLYING BY 9

DIRECTIONS: When multiplying by 9 always multiply in two steps. First multiply the number by 10. Then subtract the original number from that product, and you will have the answer.

Example A: 15 x 9

Step 1:	Multiply the 15 by 10	15 X 10 = 150
Step 2:	Subtract 15 from 150	150 − 15 = 135
Answer:	135	

****

Example B: 24 x 9

Step 1:	Multiply the 24 by 10	24 X 10 = 240
Step 2:	Subtract 24 from 240	240 − 24 = 216
Answer:	216	

Practice Problems:

1. 13 x 9 =
2. 14 x 9 =
3. 18 x 9 =
4. 22 x 9 =
5. 25 x 9 =

6. 30 x 9 =
7. 35 x 9 =
8. 38 x 9 =
9. 40 x 9 =
10. 150 x 9 =

Lesson 5 MULTIPLYING BY 11

Directions: When multiplying by 11, first write down any two-digit number and leave a space in the middle. Place the first number in the hundreds columns and the second number in the ones column. Secondly, add the two numbers together and place that number in the middle of the two beginning numbers or in the tens column.

Example A: 53 x 11

Hundreds/Tens/Ones ⟶ H /T /O

Step 1: Write the number 53 down and leave a space in the middle.

Hundred	Ten	Ones
5	—	3

Step 2: Find the sum of 5 and 3 by adding. $5 + 3 = 8$

Step 3: Place the sum (8) between the 5 and 3. 583

Hundred	Ten	Ones
5	<u>8</u>	3

Answer: 583

****

Example B: 67 x 11 (This time you must carry the tens to the hundreds column)
Hundreds/Tens/One   H/ T /0

Step 1: Write the number down and leave a space in the middle. 6 _ 7

Step 2: Find the sum of 6 and 7 by adding. $6 + 7 = 13$

Step 3: Place the 3 in the middle and add the one to the 6.

$^{7}6_{\backslash}37 = 737$

Answer: 737

Practice Problems:

1. 32 x 11 =	6. 72 x 11 =
2. 44 x 11 =	7. 61 x 11 =
3. 54 x 11 =	8. 19 x 11 =
4. 14 x 11 =	9. 55 x 11 =
5. 25 x 11 =	10. 38 X 11 =

Lesson 6 MULTIPLYING BY 12

DIRECTIONS: First, you must always split the 12 to 10 and 2. Then multiply the original number by 10. Next, you must multiply the original number by 2. Finally, add the two products together.

Example A:	15 x 12	
Step 1:	Split the 12 to 10 add 2	10 + 2 = 12
Step 2:	Multiply the 15 x 10	15 x 10 = 150
Step 3:	Multiply the 15 x 2	15 x 2 = 30
Step 4:	Add 150 + 30 = 180	
Answer:	180	

****

Example B:	14 x 12	
Step 1:	Split the 12 to 10 and 2	10 + 2 = 12
Step 2:	Multiply the 14 x 10	14 x 10 = 140
Step 3:	Multiply the 14 x 2	14 x 2 = 28
Step 4:	Add 140 + 28 = 168	
Answer:	168	

Practice Problems:

1.	16 x 12 =	6.	35 x 12 =
2.	13 x 12 =	7.	36 x 12 =
3.	24 x 12 =	8.	40 x 12 =
4.	27 x 12 =	9.	45 x 12 =
5.	30 x 12 =	10.	50 x 12 =

Lesson 7 MULTIPLYING BY NUMBERS THAT END IN ½ .

DIRECTIONS: Double the number ending in . and cut the whole number in half.

Example **A:**	14 x 2 ½.	
Step 1:	Double the 2 ½ by multiplying by 2	2 ½ x 2 = 5
Step 2:	Cut the 14 in half by dividing by 2	14 ÷ 2 = 7
Step 3:	Multiply the 5 by 7	5 x 7 = 35
Answer:	35	

****

Example **B:**	16 x 3½.	
Step 1:	Double the 3 ½ by multiplying by 2	3½ x 2 = 7
Step 2:	Cut the 16 in half by dividing by 2	16 ÷ 2 = 8
Step 3:	Multiply the 7 by 8	7 x 8 = 56

The answer is **56.**

Practice Problems:

1. 8 x 1½	6. 8 x ½ =
2. 6 x 3½	7. 12 x 4 ½ =
3. 4 x 4½	8. 14 x 5 ½ =
4. 10 x 5½	9. 16 x 2 ½ =
5. 12 x 6½	10. 18 x 4 ½ =

Lesson 8 MULTIPLYING BY SPLITTING NUMBERS

DIRECTIONS: When multiplying two numbers and at least one of the numbers has two digits, always split the larger number into two parts by dividing it by 2.

Example A: 7 x 14

Step 1:   Cut the 14 into two parts by dividing it by two. $14 \div 2 = 7$
                    Rewrite 14 by showing 7 x 2 = 14

Step 2:   Rewrite the original problem ⟶ 7 x 14 is rewritten as 7 x 7 x 2

Step 3:   Multiply 7 x 7 = 49 then, 49 x 2 = 98

Answer: 98

****

Example B: 9 x 16

Step 1:   Cut the 16 into two parts by dividing it by two. $16 \div 2 = 8$

                    Rewrite 16 by showing 8 x 2 = 16

Step 2:   Rewrite the original problem ⟶ 9 x 16 is written as 9 x 8 x 2

Step 3:   Multiply 9 x 8 = 72 then, 72 x 2 = 144

Answer: 144

Practice Problems:

1. 5 x 22 =   6. 8 x 22 =
2. 9 x 12 =   7. 9 x 14 =
3. 8 x 16 =   8. 7 x 20 =
4. 12 x 22 =   9. 4 x 32 =
5. 12 x 14 =   10. 6 x 16 =

Lesson 9 DIVIDING BY 5

DIRECTIONS: Double the whole number, then divide the product by 10.

Example A: 32 ÷ 5

Step 1:	Double the 32 by multiplying by 2	32 x 2 = 64
Step 2:	Double the 5 by multiplying by 2	5 x 2 = 10
Step 3:	Divide 64 by 10	64 ÷ 10 = 6.4

Answer: 6.4

****

Example B: 43 ÷ 5

Step 1:	Double 43 by multiplying by 2	43 x 2 = 86
Step 2:	Double the 5 by multiplying by 2	5 x 2 = 10
Step 3:	Divide 86 by 10	86 ÷ 10 = 8.6

Answer: 86

Practice Problems

1.	18 ÷ 5 =	6.	24 ÷ 5 =
2.	70 ÷ 5 =	7.	36 ÷ 5 =
3.	84 ÷ 5 =	8.	54 ÷ 5 =
4.	40 ÷ 5 =	9.	28 ÷ 5 =
5.	22 ÷ 5 =	10.	34 ÷ 5 =

Lesson 10 DIVIDING NUMBERS THAT END IN ½.

DIRECTIONS: Double the whole number, then double the number ending in ½

Example A: $30 \div 2\frac{1}{2}$

Step 1:	Double the 30 by multiplying by 2	$30 \times 2 = 60$
Step 2:	Double the 2½ by multiplying by 2	$2\frac{1}{2} \times 2 = 5$
Step 3:	Divide the 60 by the 5	$60 \div 5 = 12$

Answer: 12

****

Example B: $27 \div 4\frac{1}{2}$

Step 1:	Double the 27 by multiplying by 2	$27 \times 2 = 54$
Step 2:	Double the 4 ½ by multiplying by 2	$4\frac{1}{2} \times 2 = 9$
Step 3:	Divide the 54 by the 9	$54 \div 9 = 6$

Answer: 6

Practice Problems:

1.	$40 \div 2. =$	6.	$36 \div 4. =$
2.	$33 \div 1. =$	7.	$65 \div 6. =$
3.	$24. \div 3. =$	8.	$14 \div 3. =$
4.	$20 \div 2. =$	9.	$50 \div 2. =$
5.	$44 \div 5. =$	10.	$7. \div 1. =$

Directions: When squaring a number that ends in 5, like 25 x 25, always get the right side of the answer first. It should always be 25, because 5 x 5 is 25, that's a given. To get the left side of the answer, always add one to the left number and multiply it by the original number.

Example A: 35 x 35

Step 1:　　Get the right side of the answer by always

multiplying 5 x 5 = 25 ——————→ Right side answer

5 x 5 = 25

Step 2:　　Add one (1) to the 3(3 + 1 = 4) and multiply

time 3 ——————→ Left side answer

3 x 4 = 12

Step 3:　　Now write the 12 and place the 25 beside it 12 25

Answer:　　1225

****

Example B: 45 x 45

Step 1:　　Get the right side of the answer by always

multiplying 5 x 5 = 25 ——————→ Right side answer

5 x 5 = 25

Step 2:　　Add one (1) to the 4 (4 + 1 = 5) and multiply

times ——————→ Left side answer

4 x 5 = 20

Step 3:　　Now write the 20 and place the 25 beside it 20 25

Answer:　　2025

Practice Problems:

1.	15 x 15 =		6.	25 x 25 =
2.	55 x 55 =		7.	95 x 95 =
3.	65 x 65 =		8.	105 x 105 =
4.	75 x 75 =		9.	115 x 115 =
5.	85 x 85 =		10.	125 x 125 =

Test Percentage: 4 pts for each

1.	100	A 90 – 100%
2.	12	B 80- 89%
3.	9.4	C 70 – 79%
4.	144	D 60 – 69%
5.	5	F 59% or below
7.	20	
8.	300	
9.	7	
10.	180	
11.	.64	
12.	9	
13.	10	
14.	8	
15.	400	
16.	40	
17.	20	
18.	125	
19.	625	
20.	18	
21.	16	
22.	20	
23.	10 ¼	
24.	$16.69	
25.	40	

$$7 \times 7 + 1 + 50 - 25 =$$

$$8 \times 8 + 36 + 60 =$$

PHASE I - MENTAL MATH
Mental Math Complete Pages 2-52

Assessment Test

ESTIMATE

1. 36 + 20 =	2. 54 + 46 =	3. 29 + 21 =
4. 81 + 19 =	5. 72 + 28 =	6. 38 + 64 =
7. 58 – 19 =	8. 75 + 16 – 8 =	9. 15 + 27 – 10 =

SOLVE BY USING THE SHORT CUT METHODS

1. 38 + 69 + 56 + 35 =	2. 38 x 4 =	3. 64 ÷ 4 =
4. 63 x 11 =	5. 34 x 11 =	6. 47 x 11 =
7. 36 x 12 =	8. 12 x 4. =	9. 6 x 3. =
10. 18 x 5 =	11. 30 x 5 =	12. 41 x 5 =
13. 24 ÷ 5 =	14. 32 ÷ 5 =	15. 27 ÷ 5 =
16. 35 x 35 =	17. 65 x 65 =	18. 85 x 85 =

Mental Math Problems

1. 3 x 8 + 1 + 75 =

2. 9 x 9 + 19 =

3. 8 x 8 + 36 =

4. 7 x 7 + 1 – 8 =

5. 12 x 12 – 44 ÷ 10 =

6. 9 x 6 + 44 + 21 – 1 + 25 =

7. 3 x 9 + 3 – 10 X 5 + 100 ÷ 25 =

8. 12 x 12 + 6 – 25 – 5 ÷ 12 =

9. 6 x 8 + 2 + 50 + 100 + 100 ÷ 25

10. 1/2 0f 50 x 5 + 75 – 50 ÷10 =

11. 1/3 of 27 x 9 + 19 + 100 – 75 =

12. 1/4 of 100 + 75 + 21 + 79 =

13. 9 x 9 + 19 + 100 + 100 ÷ the # of seconds in a minute =

14. 12 x 12 + 6 – 29 take a square root x 5 + 45 – 5 ÷ 10 =

15 Two legs on a rectangle are 8 and 9. Find the area, + 28 + 75 – 25 =

16 The cube root of 64 X 4 + 4 x 5, cut that in half, cut that in half, – 8 =

17 The square root of 169 + 2 x 5 – 25 x 5 ÷ 25 = 18.

18 8 x 8 + 36 + 100 + 100 ÷ 25 – 2 x 10 – 4 ÷ 10 =

19. The # of ounces in a pound x 4, + 36 – 36 + 6 + 20 ÷ 10 x 9 + 19 + 96 =

20. 6 x 6 + 36 + 28 – 28 – 2 + 30 + 70 + 30 + 100 ÷ 25 + 113 – 25 + 37 =

Key: # = number

÷ = division sign

Mental Algebra

Examples: $4 + n = 17$ $\longrightarrow$ $n = 13$

$\qquad 6 + b = 11$ $\longrightarrow$ $b = 5$

$\qquad 2(a + b) = ?$ $\longrightarrow$ If $a = 3$ and $b = 4$ answer $6 + 8 = 14$

Word Problems for Mental Math:

a. Tessa went to the store and bought \$ 5.00 worth of apples. She also bought some pears. The total bill was \$ 8.00. How much money did the pears cost?

$5 + n = 8$ $8.00 - 5.00 = \$ 3.00$ the pears cost three dollars.

b. Wanda went shopping at the mall. At Burdines and Macys, she purchased a number of items. When she came home, she discovered that she had spent a total of \$ 210.00. At Macys she spent \$89.00. How much money did she spend at Burdines?

\$ 89.00 + w = \$ 89.00 \$ 210.00 - \$ 89.00 = \$ 121.00
Therefore, \$121.00 was spent at Burdines

c. Willie and Ann each purchased two books at \$ 10.00 and two tapes at \$ 20.00. How much money did they spend all together?

$2 (a + b) = ?$ $2(10.00 + 20.00) = 20.00 + 40.00 = 60.00$

Practice Problems

1. Al had twenty dollars. He purchased vegetables for \$8.35 and received \$8.20 back from the cashier. How much money did Al spend for the other items he purchased?

2. Chase purchased three mental math tapes for \$12.00 each, and three t-shirts for \$15 each. How much did he spend altogether?

3. Jesse gives you \$7 and says: "This is half of the money left in my wallet." How much money did Jesse have in his wallet before he gave you the \$7?

Math Game # 1

<u>Add numbers diagonally, vertically and horizontally to = 15</u> Place the numbers 1 – 9 into the 3 x 3 grid below. When you add the numbers vertically, horizontally and diagonally, all sides should equal to 15. Use the numbers only once: 1,2,3,4,5,6,7, 8, 9

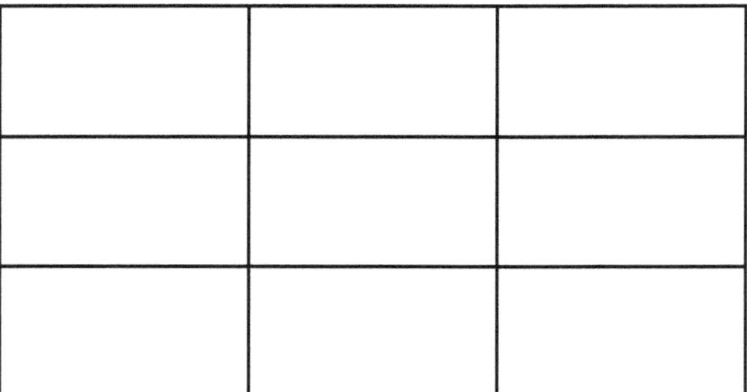

Math Game # 2

<u>Connect nine with four lines</u>
Draw four lines (horizontally, vertically or diagonally) connect all nine numbers, without lifting your pencil from your paper! You are allowed to move 4 times.

1	2	3
4	5	6
7	8	9

Math Game # 3

Which Number Am I?

- The number doesn't have a 5 or 8
  - It is not divisible by 10
  - It is not divisible by 9
  - It is not divisible by 7

70	25	49	14	54
40	45	63	72	20
18	80	28	64	30
50	48	81	27	35

Math Game # 4

Make a new upside-down triangle by only moving three coins?
Needed: 10 coins
1. Place 10 coins on the table.
2. Now, moving only three coins turn this triangle upside down!

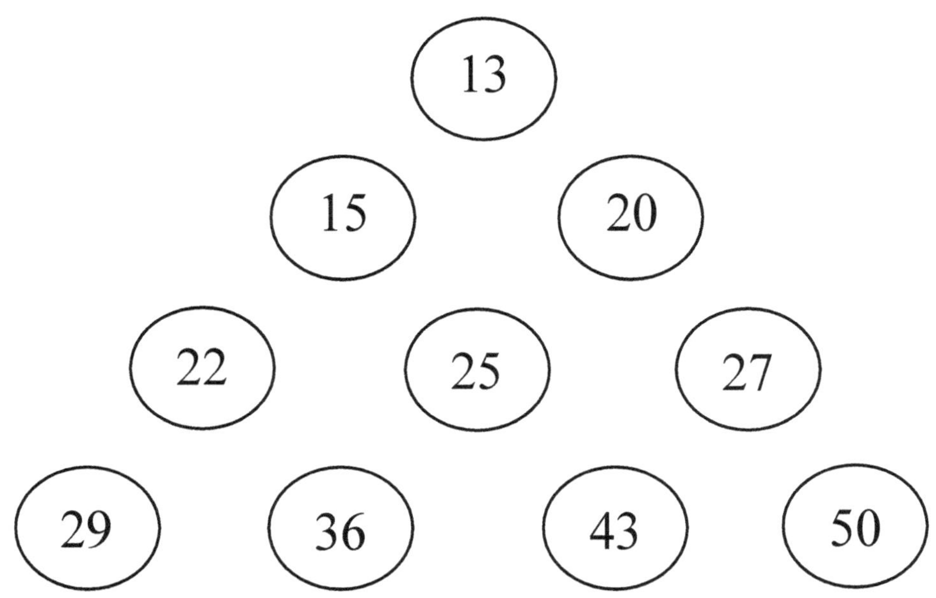

Math Tricks Game

- There are six pages with numbers from 1 to 63

- No page will exceed number 63

- Pick one of these numbers

- Let the facilitator know from page 1 through 6 if your number is listed on each page or not

- At the conclusion of page six the facilitator will tell you your number

- How to find the answer:

    a. Each page represents a number to be added mentally. Make sure the number that is given is on the following pages

    b. Page   Add
                    Mentally
        Page 1 = 1
        Page 2 = 2
        Page 3 = 4
        Page 4 = 8
        Page 5 = 16
        Page 6 = 32

    c. Have someone pick a number from 1 – 63. Next, have them tell you page by page whether the number they selected is on each page or not. If the number they have selected is on any page listed above you add these numbers mentally then tell the person what number they have selected. For example, if they select 37, they will tell you that it is on pages 1, 3, and 6.

        Page 1 = 1
        Page 3 = 4
        Page 6 = <u>32</u>

        Total = 37

Math Tricks - Page 1

1	3	5	7	9	11	13	15	17	19
21	23	25	27	29	31	33	35	37	39
41	43	45	47	49	51	53	55	57	59
61	63								

Math Tricks - Page 2

2	3	6	7	10	11	14	15	18	19
22	23	26	27	30	31	34	35	38	39
42	43	46	47	50	51	54	55	58	59
62	63								

Math Tricks - Page 3

4	5	6	7	12	13	14	15	20	21
22	23	28	29	30	31	36	37	37	39
44	45	46	47	52	53	54	55	60	61
62	63								

Math Tricks - Page 4

8	9	10	11	12	13	14	15	24	25
26	27	28	29	30	31	40	41	42	43
44	45	46	47	56	57	58	59	60	61
62	63								

Math Tricks - Page 5

16	17	18	19	20	21	22	23	24	25
26	27	28	29	30	31	48	49	50	51
52	53	54	55	56	57	58	59	60	61
62	63								

Math Tricks - Page 6

32	33	34	35	36	37	38	39	40	41
42	43	44	45	46	47	48	49	50	51
52	53	54	55	56	57	58	59	60	61
62	63								

PHASE II
Walkerdoo Math
Puzzles Instructions

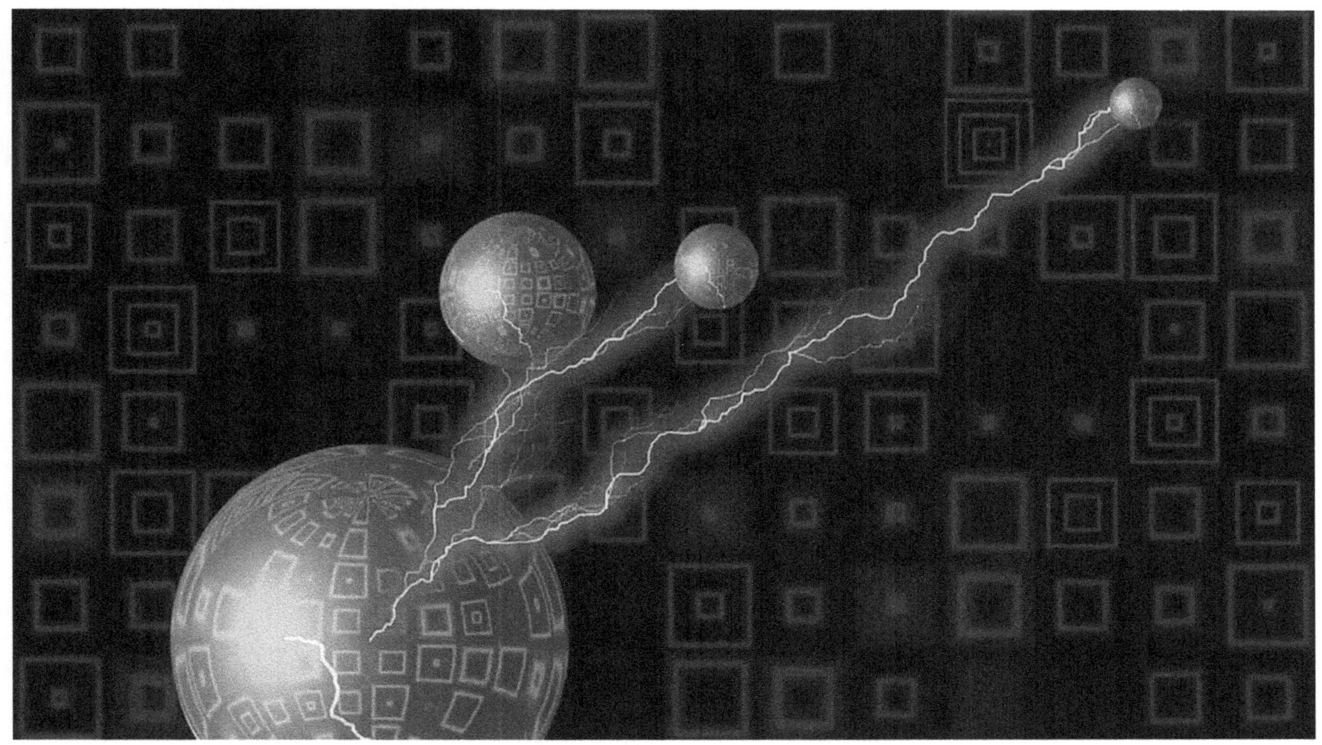

Possibly Help Fight Dementia
Volume 1

Welcome to Walkerdoo! One of the nation's leading mathematical minds has given you hands-on access to a world of intellectually stimulating puzzles and possible adventures.

Did you know that puzzles are more than just a way to pass time or give your brain a workout? They're also great for keeping your mind healthy and sharp! Research has shown that engaging in activities like the ones you will find in Walkerdoo can have some fantastic benefits, including:

1. **Cognitive Stimulation:** Solving puzzles challenges your brain and keeps it active, which can help improve cognitive function and memory (Brooker et al., 2019).
2. **Reduced Risk of Dementia:** Studies have found that regularly engaging in mentally stimulating activities, like puzzles, may help lower the risk of developing dementia and Alzheimer's disease (Corbett et al., 2024).
3. **Improved Mood:** Puzzle-solving releases dopamine, a neurotransmitter associated with pleasure and reward (The Healthline Editorial Team, 2019). Not only are you exercising your brain, but you are also increasing your mood!

Solving one of Walkerdoo's puzzles means you're not just having fun but investing in your long-term brain health.

And hey, this is just the beginning! Walkerdoo Volume 1 is the start of an incredible adventure. We have exciting plans for the future, with many more volumes filled with more brain-provoking challenges and surprises. Stay tuned for more.

So, grab your favorite pencil, cozy up with a cup of relaxing tea, and let's get started. Your brain will thank you!

How to Solve Walkerdoo Stimulating Math Puzzles

Walkerdoo is a grid-based game designed to challenge your number pattern recognition and arithmetic skills. Here's how to get started:

- Understanding the Grid:
    - The grid consists of columns (vertical lines) and rows (horizontal lines) Key Columns - A and B: Columns A and B are the foundation of the game. They contain initial numbers that establish a pattern or sequence. (You may also complete the puzzle without knowing the pattern or sequence)
- Identifying Patterns:
    - In Column A, the pattern could involve starting with the number 2 then adding 3, and then subtract 1 alternately. Each column may have a different pattern.
- Building the Grid:
    - All subsequent rows in the grid are generated based on the numbers in rows A and B. The numbers in these rows will follow the established patterns.
- Solving the Puzzle:
    - Remember to perform math operations, such as adding, subtracting, multiplying or dividing numbers, to find the correct answer for a particular **blank** cell.
- Example Puzzle: - Column A follows the pattern: start at 2, add 3, subtract 1= 4), and so on.
    - Rows A and B across have initial numbers: A: [2], x :row B [1] = [2], and so on

Walkerdoo Math Game Puzzle – Easy

Note the pattern in Columns A and B first. Then perform the math operation in each row across and fill in each blank cell

Columns:	A	B	1	2	3
Rows: ⟶		x	=	+	=
	2		2	6	
	5	2	10		18
	4	3		7	
		4	28	6	

Above Puzzle Answer Key

Columns:	A	B	1	2	3
Rows: ⟶		x	=	+	=
	2	1	2	6	8
	5	2	10	8	18
	4	3	12	7	19
	7	4	28	6	34

By following these instructions and observing the patterns, you'll be able to master Walkerdoo and enjoy the challenge it presents.

Walkerdoo Math Game Puzzle 1 – Easy

Note the pattern in Columns A and B. Then perform the math operation in each Row across and fill in each blank cell

Columns:	A	B	1	2	3
Rows: →	x	=	+	=
	3	2	6		12
	6	2		8	20
	4		8		13
	7	3		7	28
		3	15	4	
	8		24		30
	6	4		3	27
	9	4	36	5	
	7	4	28	2	

Walkerdoo Math Game Puzzle 2 – Easy

Note the pattern or sequence in Columns A or B. Then perform the
arithmetic operation in each Row and fill in each blank

Columns:	A	B	1	2	3
Rows: →		x	=	-	=
	6	8	48		41
	4		20	10	
	7	7		8	41
	5	4	20		9
	8		48	9	
		3	18		6
	9		45		35
	7	2		13	
	10	4	40		29

Walkerdoo Math Game Puzzle 3 – Easy

Note the pattern or sequence in Columns A or B. Then perform the math operation in each row across and fill in each blank

Columns:	A	B	1	2	3
Rows: →		x	=	-	=
	4	4		6	10
	2		4		1
	5		25	5	
		3	9		7
	6		36	12	
	4		16		11
		7	49	10	
	5		25		17
	8	8		9	

Walkerdoo Math Game Puzzle 3 – Easy

Walkerdoo Math Game Puzzle 4 – Easy

Note the pattern or sequence in Columns A or B. Then perform the math operation in each Row across and fill in each blank

Columns:	A	B	1	2	3
Rows: →		+	=	-	=
	1^2	5	6		4
	2^2	8	12	6	
	3^2	6		7	8
	4^2	9	25		5
	5^2		32	7	
	6^2	10	46		30
	7^2		57	7	
		11	75		65
	9^2		90	15	

Note: To square a number means to multiply the number by itself. For example, 5^2, is 5 x 5 equals 25.

Walkerdoo Math Game Puzzle 5 – Easy

Note the pattern or sequence in Columns A or B. Then perform the math operation in each Row across and fill in each blank

Columns:	A	B	1	2	3
Rows: →		+	=	x	=
	2		11		33
	5	8		4	52
	3		10		50
	6		12		72
	4	5		7	
		4	11		88
	5		8		72
	8		10	10	
	6	1	7		77

Walkerdoo Math Game Puzzle 6 – Easy

Note the pattern or sequence in Columns A or B. Then perform the
math operation in each Row across and fill in each blank

Columns:	A	B	1	2	3
Rows: ⟶		x	=	-	=
	1	9		5	4
	4		32		28
	2	7	14	3	
	5		30		28
		5	15		14
	6			2	22
	4	3	12		9
	7	2		4	10
	5			5	0

Walkerdoo Math Game Puzzle 7 – Easy

Note the pattern or sequence in Columns A or B. Then perform the math operation in each Row across and fill in each blank

Columns:	A	B	1	2	3
Rows: →		x	=	÷	=
		3	30		10
	9		27	3	
	8	3		3	
		3	21		7
	6	3		3	6
	5		15		5
		3	12	3	
	3	3		3	
	2		6		2

Walkerdoo Math Game Puzzle 8 – Easy

Note the pattern or sequence in Columns A or B. Then perform the
math operation in each Row across and fill in each blank

Columns:	A	B	1	2	3
Rows: ⟶		x	=	+	=
	2	1	2	9	
	3		6		14
	4	3		7	19
	5		20		26
	6	5		5	
		6	42		46
	8		56	3	
	9	8			74
		9	90	1	

Walkerdoo Math Game Puzzle 9 – Easy

Note the pattern or sequence in Columns A or B. Then perform the
math operation in each Row across and fill in each blank

Columns:	A	B	1	2	3
Rows: →		x	=	÷	=
	6	11	66		11
	6	10		6	10
	6		54		9
	6		48	6	
		7	42		7
	6	6		6	
	6	5		6	
		4	24	6	
	6		18		3

Walkerdoo Math Game Puzzle 10 – Easy

Note the pattern or sequence in Columns A or B. Then perform the
math operation in each Row across and fill in each blank

Columns:	A	B	1	2	3
Rows: →		+	=	+	=
	10	1	11	9	
	9	3		8	20
		2	10		17
	7		11	6	
	6		9		14
		5	10	4	
	4		8		11
	3	6		2	
	2		7		8

Walkerdoo Math Game Puzzle 11 – Easy

Note the pattern or sequence in Columns A or B. Then perform the
math operation in each Row across and fill in each blank

Columns:	A	B	1	2	3
Rows: →	x	=	+	=	
	6	1		14	
	3		6		19
		3	15		27
	2		8	11	
	4	5		10	
	1		6		15
	3		21	8	
	0	8		7	7
	2		18		24

Walkerdoo Math Game Puzzle 12 – Easy

Note the pattern or sequence in Columns A or B. Then perform the
math operation in each Row across and fill in each blank

Columns:	A	B	1	2	3
Rows: →		x	=	-	=
	8	9	72		63
	5	8		8	
	7		49		42
		6	24		18
	6		30	5	
	3	4		4	
		3	15		12
	2		4		2
	4		4		3

Walkerdoo Math Game Puzzle 13 – Easy

Note the pattern or sequence in Columns A or B. Then perform the
math operation in each Row across and fill in each blank

Columns:

	A	B	1	2	3
Rows: →		+	=	+	=
	18	12	30	2	
	16	11	27		31
	14		24	6	30
		9	21		29
	10	8		10	
	8			12	27
		6	12		26
	4		9		25
	2		6	18	

Walkerdoo Math Game Puzzle 14 – Easy

Note the pattern or sequence in Columns A or B. Then perform the math operation in each Row across and fill in each blank

Columns:	A	B	1	2	3
Rows: ⟶			÷	=	x	=
	12	6	2		24
	10		2	11	
	14		2		20
		6	2		18
	16	8		8	
	14		2		14
		9		6	12
	16		2		10
	20	10			8

Walkerdoo Math Game Puzzle 15 – Easy

Note the pattern or sequence in Columns A or B. Then perform the math operation in each Row across and fill in each blank

Columns:	A	B	1	2	3
Rows: →		÷	=	x	=
	81	9		2	
	72		8		40
		9	7		28
	54		6		42
		9	5	6	
	36	9		9	
		9	3		24
	18	9			22
		9	1		10

Walkerdoo Math Game Puzzle 16 – Easy

Note the pattern or sequence in Columns A or B. Then perform the
math operation in each Row across and fill in each blank

Columns:	A	B	1	2	3
Rows: →		+	=	x	=
	9^2	9	90		90
	6^2		44		88
	8^2	7		1	71
	5^2	6		2	
		5	54	1	
	4^2		20		40
	6^2	3			39
	3^2		11		22
	5^2	1			26

Note: To square a number means to multiply the number by itself. For example, 5^2, is 5 x 5
equals 25.

Walkerdoo Math Game Puzzle 17 – Easy

Note the pattern or sequence in Columns A or B. Then perform the math operation in each Row across and fill in each blank

Columns:	A	B	1	2	3
Rows: ⟶		x	=	÷	=
	4	11	44		11
	4		40	4	
		9	36		9
	4		32	4	
	4	7		4	7
		6	24		6
	4		20	4	
		4	16		4
	4		12	4	

Walkerdoo Math Game Puzzle 18 – Easy

Note the pattern or sequence in Columns A or B. Then perform the math operation in each Row across and fill in each blank

Columns:	A	B	1	2	3
Rows: →		x	=	+	=
	1		9		13
	4	8		2	34
	2	7	14		20
	5		30	4	
	3		15		23
		4	24		30
	4		12	10	
		2	14		22
	5	1		12	

Walkerdoo Math Game Puzzle 19 – Easy

Note the pattern or sequence in Columns A or B. Then perform the math operation in each Row across and fill in each blank

Columns:	A	B	1	2	3
Rows: ⟶		+	=	x	=
		5	6		48
	2		10	5	
	3	4		7	
	4		11		44
		3	8		48
	6		12	3	
	7		9		45
	8	5		2	26
	9		10	4	

Walkerdoo Math Game Puzzle 20 – Easy

Note the pattern or sequence in Columns A or B. Then perform the
math operation in each Row across and fill in each blank

Columns:	A	B	1	2	3
Rows: →		÷	=	x	=
		2	15		45
	27	3	9		27
		2	12	3	
	21		7		28
		2	9		36
	15		5	4	
	12	2		5	
	9	3		5	
		2	3	5	

Walkerdoo Math Game Puzzle 21 – Easy

Note the pattern or sequence in Columns A or B. Then perform the
math operation in each Row across and fill in each blank

Columns:	A	B	1	2	3
Rows: →		x	=	-	=
	2	9		9	
	2		16		7
	3	7		7	14
		6	18		11
	4		20	5	
	4	4		5	11
	5		15		12
	5		10	3	
	6	1	6		5

Walkerdoo Math Game Puzzle 22 – Easy

Note the pattern or sequence in Columns A or B. Then perform the
math operation in each Row across and fill in each blank

Columns:	A	B	1	2	3
Rows: →		+	=	-	=
	3		18		16
	3	12		3	
	4		17		15
		10	14		11
	5	11		2	
	5	8		3	
		9	15		13
	6		12	3	
		7	14		12

Walkerdoo Math Game Puzzle 23 – Easy

Note the pattern or sequence in Columns A or B. Then perform the math operation in each Row across and fill in each blank

Columns:	A	B	1	2	3
Rows: →		+	=	-	=
	4		9		6
		1	5	3	
	5	7		3	
	5		8		4
		9	15	4	
	6		11		7
	7		18		13
	7	7		5	
		13	20		15

Walkerdoo Math Game Puzzle 24 – Easy

Note the pattern or sequence in Columns A or B. Then perform the math operation in each Row across and fill in each blank

Columns:	A	B	1	2	3
Rows: →		+	=	-	=
	1^3	9	10		4
	1^3		9	6	
	1^3	7		6	2
	2^3		14	8	6
	2^3		13		5
	2^3	4		8	4
	3^3	3	30		20
	3^3		29		19
	3^3	1	28	10	

Note: To cube a number means to multiply the number by itself three times. For example, 3^3 or 3 cubed means, $3 \times 3 \times 3 = 27$.

Walkerdoo Math Game Puzzle 25 – Easy

Note the pattern or sequence in Columns A or B. Then perform the
math operation in each Row across and fill in each blank

Columns:	A	B	1	2	3
Rows: →		x	=	+	=
	5	1		15	20
	7		14		28
	4	3		13	25
	6		24	12	
		5	15		26
	5	6		10	
	2		14		23
		8	32	8	
	1	9		7	16

Walkerdoo Math Game Puzzle 26 – Easy

Note the pattern or sequence in Columns A or B. Then perform the
math operation in each Row across and fill in each blank

Columns:	A	B	1	2	3
Rows: →		x	=	+	=
	1	9		12	
	3		24	11	
	2	7		10	24
	4		24		33
	3	5		8	23
		4	20		27
	4		12	6	
		2	12	5	
	5	1		4	9

Walkerdoo Math Game Puzzle 27 – Easy

Note the pattern or sequence in Columns A or B. Then perform the
math operation in each Row across and fill in each blank

Columns:	A	B	1	2	3
Rows: →		x	=	÷	=
	12		72		12
	11	6		6	11
		6	60	6	
		6	54	6	
	8		48		8
	7	6		6	7
		6	36	6	
	5	6		6	
		6	24		4

Walkerdoo Math Game Puzzle 28 – Easy

Note the pattern or sequence in Columns A or B. Then perform the
math operation in each Row across and fill in each blank

Columns:	A	B	1	2	3
Rows: ⟶		+	=	+	=
	4	1	5		23
		4	8	20	
	4		7		23
		6	11		29
	5		10	14	
	5		13		29
		7	13	12	
	6		16		30
	6		15	10	

Walkerdoo Math Game Puzzle 29 – Easy

Note the pattern or sequence in Columns A or B. Then perform the math operation in each Row across and fill in each blank

Columns:	A	B	1	2	3
Rows: ⟶		+	=	-	=
	5^2	8	33		24
	5^2	7		8	24
	4^2	22			15
	4^2		21	6	
	3^2	4		5	8
	3^2		12		8
	2^2		6	3	
	2^2	1		2	3
	1^2		1		0

Note: To square a number means to multiply the number by itself. For example, 5^2, is 5 x 5 equals 25.

Walkerdoo Math Game Puzzle 30 – Easy

Note the pattern or sequence in Columns A or B. Then perform the
math operation in each Row across and fill in each blank

Columns:

	A	B	1	2	3
Rows: →		x	=	+	=
1	1		4		12
2	2	7		10	24
3		5		7	22
4	4	8		9	41
5		6	30		36
6	6		54	8	
7	7		49		54
8	8		80	7	
9		8	72		76

Walkerdoo Math Game Puzzle 31 – Easy

Note the pattern or sequence in Columns A or B. Then perform the math operation in each Row across and fill in each blank

Columns:	A	B	1	2	3
Rows: →		÷	=	x	=
	30	2		10	150
	27		9		81
		2	12		96
	21	3		7	
		2	9		54
	15		5	5	
		2	6		24
	9		3	3	
	6		3		6

Walkerdoo Math Game Puzzle 32 – Easy

Note the pattern or sequence in Columns A or B. Then perform the
math operation in each Row across and fill in each blank

Columns:	A	B	1	2	3
Rows: →		+	=	x	=
	9	0		4	36
	7		8	4	
	10		12		48
	8		11	3	
		4	15	3	
	9		14		42
	12		18		36
		7	17	2	
	13		21		42

Walkerdoo Math Game Puzzle 33 – Easy

Note the pattern or sequence in Columns A or B. Then perform the
math operation in each Row across and fill in each blank

Columns:	A	B	1	2	3
Rows: →		+	=	-	=
	1^2	7	8		3
	2^2		8	6	
	3^2		15		8
	4^2	3		8	11
	5^2		30		21
	6^2	2		10	28
	7^2		53		42
	8^2	1		12	53
	9^2	3	84		71

Note: To square a number means to multiply the number by itself. For example, 5^2, is 5 x 5
equals 25.

Walkerdoo Math Game Puzzle 34 – Easy

Note the pattern or sequence in Columns A or B. Then perform the
math operation in each Row across and fill in each blank

Columns:	A	B	1	2	3
Rows: →		x	=	÷	=
	2		18	2	
		8	24		8
	4		28	2	
	5		30		10
		5	30	2	
	7		28		14
	8		24	3	
		2	18		6
	10		10	2	

Walkerdoo Math Game Puzzle 34 – Easy

Walkerdoo Math Game Puzzle 35 – Easy

Note the pattern or sequence in Columns A or B. Then perform the math operation in each Row across and fill in each blank

Columns:	A	B	1	2	3
Rows: ⟶		x	=	÷	=
	6		36		6
		8	64	8	
	5		25		5
		7		7	7
	4		16	4	
	6		36		6
		3		3	3
	5		25		5
	2		4	2	

Walkerdoo Math Game Puzzle 36 – Easy

Note the pattern or sequence in Columns A or B. Then perform the math operation in each Row across and fill in each blank

Columns:	A	B	1	2	3
Rows: →		+	=	-	=
	9	2^2	13		6
		2^2	12	7	
	7	3^2		7	9
		3^2	15		10
	5	4^2		5	16
	4		20	5	
		5^2	28		25
		5^2	27	3	
	1	6^2		3	34

Note: To square a number means to multiply the number by itself. For example, 5^2, is 5 x 5 equals 25.

Walkerdoo Math Game Puzzle 37 – Easy

Note the pattern or sequence in Columns A or B. Then perform the
math operation in each Row across and fill in each blank

Columns:	A	B	1	2	3
Rows: ⟶		x	=	÷	=
	12		84		12
		7	77	7	
	10		70		10
	9	7		7	
		7	56		8
	7		49		7
		7	42	7	
	5	7		7	
		7	28	7	

Walkerdoo Math Game Puzzle 38 – Easy

Note the pattern or sequence in Columns A or B. Then perform the
math operation in each Row across and fill in each blank

Columns:

	A	B	1	2	3
Rows: →			-	=	x	=
	7	2	5	9	
	8		4		32
		3	6	7	
	10		5		30
		4	7		35
		6	6	4	
	13	5		3	24
		7	7		14
	15		9	1	

Walkerdoo Math Game Puzzle 39 – Easy

Note the pattern or sequence in Columns A or B. Then perform the math operation in each Row across and fill in each blank

Columns: A	B	1	2	3
Rows: →	+	=	-	=
10	2^3		9	9
12	2^3	20		12
	2^3	17		10
	3^3	38	6	
8	3^3		5	
	3^3	37		33
	4^3	71`	3	
9	4^3		2	71
6	4^3		1	69

Note: To cube a number means to multiply the number by itself three times. For example, 3^3 or 3 cubed means, 3 x 3 x 3 = 27.

Walkerdoo Math Game Puzzle 40 – Easy

Note the pattern or sequence in Columns A or B. Then perform the math operation in each Row across and fill in each blank

Columns:	A	B	1	2	3
Rows: →		x	=	+	=
	1	2		12	
		2	4		12
	3		6		16
		3		6	18
	5		15	8	
		3	18		22
		4	28		34
	8	4		2	34
		4	36		40

Walkerdoo Math Game Puzzle 41 – Easy

Note the pattern or sequence in Columns A or B. Then perform the
math operation in each Row across and fill in each blank

Columns:	A	B	1	2	3
Rows: →		+	=	x	=
		15	24		48
	8		20	3	
		13		2	40
	6		16		48
	5	11		2	32
		8	12	3	
	3		12		24
		6	8	3	
	1		8		16

Walkerdoo Math Game Puzzle 42 – Easy

Note the pattern or sequence in Columns A or B. Then perform the
math operation in each Row across and fill in each blank

Columns:	A	B	1	2	3
Rows: ⟶			÷	=	x	=
	80		10		50
		8	9	4	
	64	8		7	
	56		7		42
		8	6	9	
		8	5		40
	32		4	11	
		8	3		30
		8	2		26

Walkerdoo Math Game Puzzle 43 – Easy

Note the pattern or sequence in Columns A or B. Then perform the
math operation in each Row across and fill in each blank

Columns:	A	B	1	2	3
Rows: →		x	=	+	=
		1	5	12	
	5	4		8	28
	5	3	15		25
		6	36		42
	6		30	8	
	6		48		52
		7	49		55
	7		70	2	
	7		63		67

Walkerdoo Math Game Puzzle 44 – Easy

Note the pattern or sequence in Columns A or B. Then perform the math operation in each Row across and fill in each blank

Columns:	A	B	1	2	3
Rows: $\longrightarrow$		+	=	-	=
	5^2	5		15	15
	5^2		33	6	
	5^2	6		7	24
	4^2		25		5
	4^2		23	7	
	4^2	10		16	10
	3^2		17		9
	3^2	11		10	
	3^2		18		15

Note: To square a number means to multiply the number by itself. For example, 5^2, is 5 x 5 equals 25.

Walkerdoo Math Game Puzzle 45 – Easy

Follow the pattern or sequence in Columns A or B. Then perform
the arithmetic operation in each Row and fill in each blank

Columns:	A	B	1	2	3
Rows: →		x	=	-	=
	9		45	9	
		8	64		56
	7		28		21
		7	42	6	
	5		15		10
	4	6		4	
		2	6		3
	2		10	2	
	1	1	1	1	

Walkerdoo Math Game Puzzle 46 – Easy

Note the pattern or sequence in Columns A or B. Then perform the math operation in each Row across and fill in each blank

Columns:

	A	B	1	2	3
Rows: →		+	=	-	=
	9^2	2	83		74
	6^2	3		8	31
	8^2		68	7	
	5^2	5		6	24
	7^2		55		50
	4^2	7	23	4	
	6^2		44		41
	3^2	9		2	16
	5^2		35	1	

Note: To square a number means to multiply the number by itself. For example, 5^2, is 5 x 5 equals 25.

Walkerdoo Math Game Puzzle 47 – Easy

Note the pattern or sequence in Columns A or B. Then perform the
math operation in each Row across and fill in each blank

Columns:	A	B	÷	1	=	2	-	3	=
			÷		=		-		=
	36			18				14	
		3		11		4			
	30	2				4		11	
	27			9				6	
		2		12		3			
	21			7				4	
		2		9		2			
	15	3						3	
	12			6		2			

Walkerdoo Math Game Puzzle 48 – Easy

Follow the pattern or sequence in Columns A or B. Then perform
the arithmetic operation in each Row and fill in each blank

Columns:	A	B	1	2	3
Rows: →		x	=	-	=
	8		40	9	
		8	48		40
	9		54	7	
	7	9		6	57
		7	70		65
		10	80	4	
	11		88		85
	9	11		2	97
		9	108	1	

Walkerdoo Math Game Puzzle 49 – Easy

Note the pattern or sequence in Columns A or B. Then perform the math operation in each Row across and fill in each blank

Columns:	A	B	1	2	3
Rows: $\longrightarrow$		+	=	-	=
	2^3	12	20		11
	2^3		22		14
	2^3	11		7	12
	3^3		36	6	
	3^3	6	33		28
	3^3		35		31
	4^3	5	`	3	66
	4^3		71		69
	4^3	4		1	67

Note: To cube a number means to multiply the number by itself three times. For example, 3^3 or 3 cubed means, 3 x 3 x 3 = 27.

Walkerdoo Math Game Puzzle 50 – Easy

Note the pattern or sequence in Columns A or B. Then perform the math operation in each Row across and fill in each blank

Columns:

	A	B	1	2	3
Rows: →		-	=	x	=
	6		4		36
		4	3	10	
	8		5		35
	9	5		8	32
		4	6	5	
	11		5		30
	12		7		21
		7	6	4	
	14		8		8

Walkerdoo Math Game Puzzle 51 – Easy

Note the pattern or sequence in Columns A or B. Then perform the math operation in each Row across and fill in each blank

Columns:	A	B	1	2	3
Rows: →			+	=	x	=
	12		13		26
		4	12	2	
	10		12		36
		5	11	3	
	8		11		44
		6		4	40
	6		10		50
		7	9	5	
	4		9		54

Walkerdoo Math Game Puzzle 52 – Easy

Note the pattern or sequence in Columns A or B. Then perform the
math operation in each Row across and fill in each blank

Columns:	A	B	1	2	3
Rows: →			÷	=	+	=
	42	2			9	30
	39		13		21
	36		18	7	
		3	11		17
	30	2		5	20
	27		9	4	
		2	12		15
	21		7		9
		2	9	1	

Walkerdoo Math Game Puzzle 53 – Easy

Note the pattern or sequence in Columns A or B. Then perform the
math operation in each Row across and fill in each blank

Columns:	A	B	1	2	3
Rows: →		x	=	-	=
	2		18	11	
		8	40		30
	3		21	9	
		6	36		28
	4	5		7	13
	7		28		22
	5		15	5	
		2	16		12
	6		6	3	

Walkerdoo Math Game Puzzle 54 – Easy

Note the pattern or sequence in Columns A or B. Then perform the math operation in each Row across and fill in each blank

Columns:	A	B	1	2	3
Rows: ⟶		+	=	÷	=
	81	9	90	9	
	72	9		9	9
	63		72		8
		9	63	9	
		9	54		6
	36		45		5
		9	36	9	
	18		27	9	
	9	9		9	2

Walkerdoo Math Game Puzzle 55 – Easy

Note the pattern or sequence in Columns A or B. Then perform the math operation in each Row across and fill in each blank

Columns:	A	B	1	2	3
Rows: →		x	=	÷	=
	5	4		4	5
	9		36	4	
		4	24		6
	10		50	5	
		5	35	5	
	11		55		11
	8	6		6	8
		6	72	6	
	9		54	6	

Walkerdoo Math Game Puzzle 56 – Medium

Note the pattern in Columns A and B first. Then perform the math operation in each row across and fill in each blank cell

Columns:

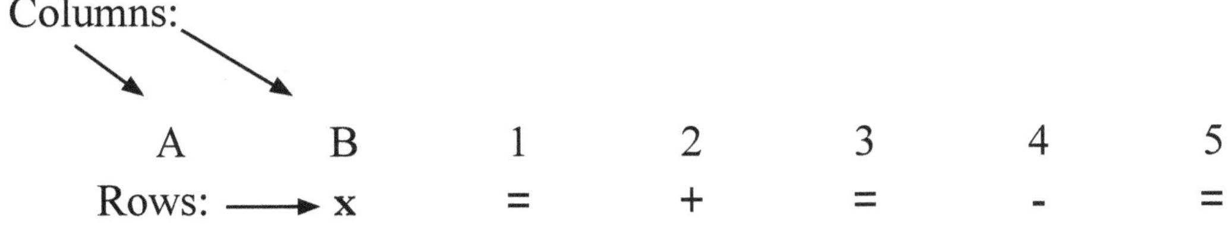

A	B	1	2	3	4	5
Rows: → x		=	+	=	-	=

A	B	1	2	3	4	5
1	8	8		13		4
2		14	10		8	
	6	18		33		26
4	5		20	40		34
	4	20	25		5	
6		18		48		44
	2	14	35		3	
8		8	40			46
9	0		45		1	

Walkerdoo Math Game Puzzle 57 – Medium

Note the pattern in Columns A and B first. Then perform the math
operation in each row across and fill in each blank cell

Columns:

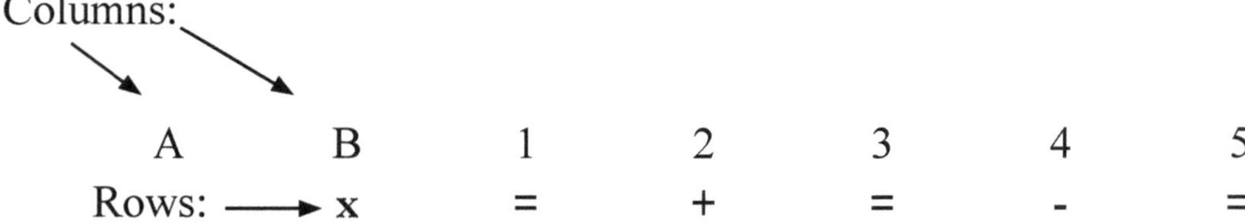

A	B	1	2	3	4	5
x		=	+	=	-	=

A	B	1	2	3	4	5
2		18		23		13
5	7		10		9	
3	8	24		39		31
6		36	20		7	49
	7	28		53	6	
7		35	30		5	
5	6		35	65	4	
	4	32		72		69
6		30	45		2	

Walkerdoo Math Game Puzzle 58 – Medium

Note the pattern in Columns A and B first. Then perform the math operation in each row across and fill in each blank cell

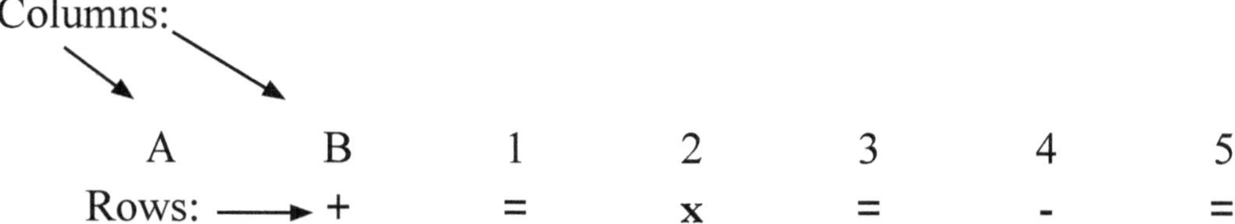

A	B	1	2	3	4	5
10	1		9		10	
9		12		96		87
	2	10	7		8	
7		11		66		59
	3	9	5		6	
5		10		40		35
	4	8		24	4	
3		9	2		3	15
2	5		1	7		5

Walkerdoo Math Game Puzzle 59 – Medium

Note the pattern in Columns A and B first. Then perform the math operation in each row across and fill in each blank cell

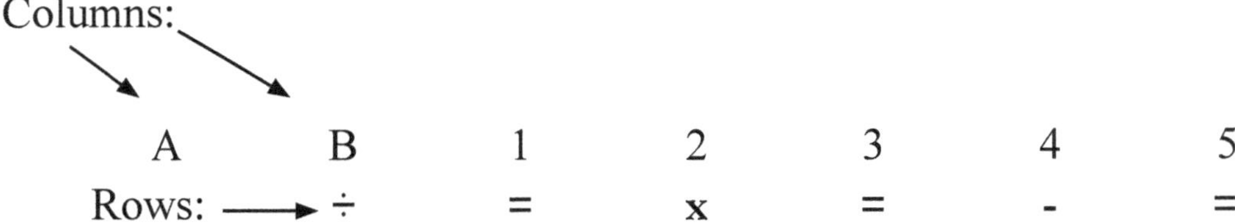

A	B	1	2	3	4	5
99	11		2		8	
88		8		40		35
	11	7		28		21
66	11		7		4	
	11	5	6		6	
44		4		36		33
	11	3		24	5	
22		2	11		2	
	11	1		10		6

Walkerdoo Math Game Puzzle 60 – Medium

Note the pattern in Columns A and B first. Then perform the math operation in each row across and fill in each blank cell

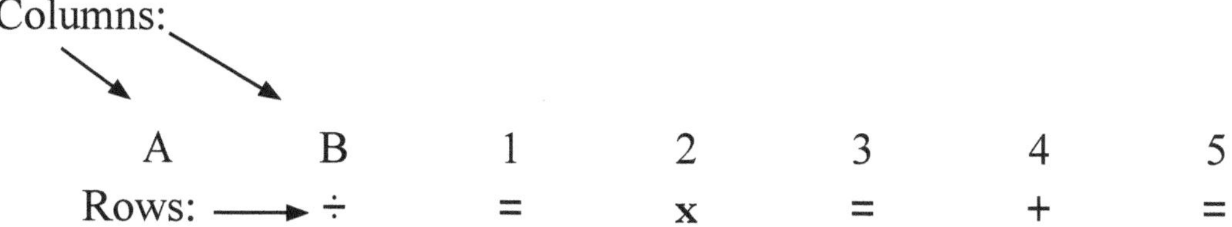

A	B	1	2	3	4	5
9^2	9	9		27		35
6^2		6	4		5	
8^2	8		5		7	47
5^2		5		30	4	
7^2		7	7		6	55
4^2	4		8	32	3	
6^2		6	9		5	59
3^2	3		10	30	2	
5^2	5	5		55	4	

Note: To square a number means to multiply the number by itself. For example, 9^2, is 9 x 9 equals 81.

Walkerdoo Math Game Puzzle 61 – Medium

Note the pattern in Columns A and B first. Then perform the math operation in each row across and fill in each blank cell

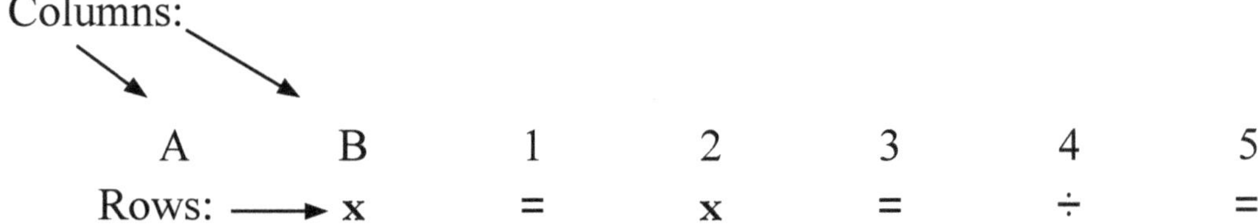

A	B	1	2	3	4	5
6	8		2	96		48
9		45	2		2	
5		35		70		35
	4	32		96		32
4	6		3	72	3	
7		21		63	3	
	5	15		60		15
6	2		4		4	
	4	8		32		8

Walkerdoo Math Game Puzzle 62 – Medium

Note the pattern in Columns A and B first. Then perform the math operation in each row across and fill in each blank cell

Columns:

	A	B	1	2	3	4	5
Rows: →	÷		=	+	=	-	=

A	B	1	2	3	4	5
36		18	12		3	
	3	11		24		21
30	2		10	25	3	
27		9		20		16
	2	12		20	4	
21	3		9	16		12
	2	9		15		10
	3	5		12	5	
12		6		10		5

Walkerdoo Math Game Puzzle 63 – Medium

Note the pattern in Columns A and B first. Then perform the math operation in each row across and fill in each blank cell

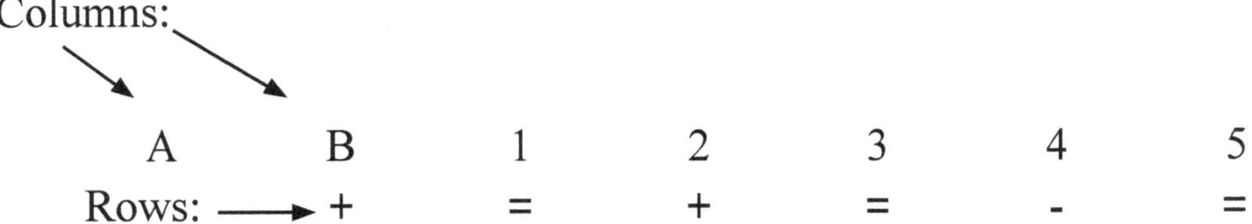

A	B	1	2	3	4	5
2^3	9	17		29	10	
2^3		16	13		9	20
	7	15		25		17
3^3		33		44	7	
3^3	5		8	40	6	
3^3		31	9		5	
4^3	3	67		73		69
4^3		66	7		3	70
4^3	1		4	69	2	

Note: To cube a number means to multiply the number by itself three times. For example, 3^3 or 3 cubed means, 3 x 3 x 3 = 27.

Walkerdoo Math Game Puzzle 64 – Medium

Note the pattern in Columns A and B first. Then perform the math operation in each row across and fill in each blank cell

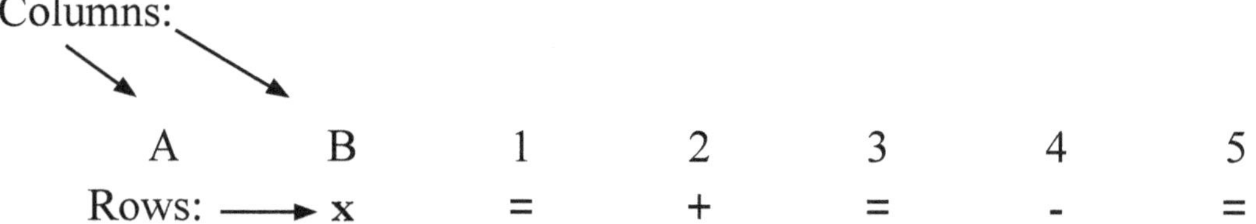

A	B	1	2	3	4	5
4	1		10	14		10
	4	16		32		25
4	3		14	26	5	
	6	30		50		42
5		25		43	6	
5		40	24		9	
	7	42		64		57
6		60	28		10	
6	9		26	80		72

Walkerdoo Math Game Puzzle 65 – Medium

Note the pattern in Columns A and B first. Then perform the math operation in each row across and fill in each blank cell

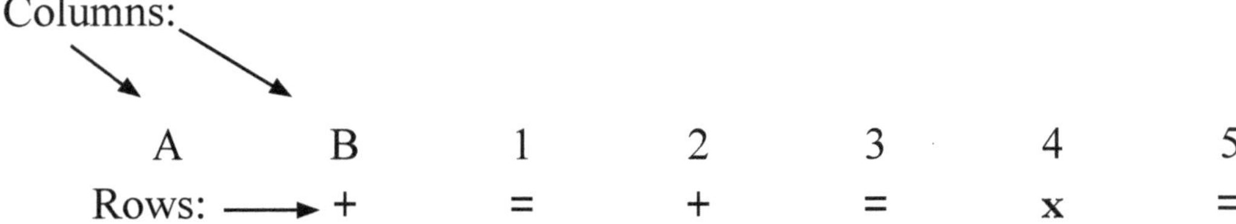

A	B	1	2	3	4	5
4	11		6		2	
7		17		24	2	
5		14	8		2	44
8	8		9		2	50
6		13		23		46
	6	15		26	2	
7		12	12		2	
	4	14		27		54
8	3		14	25	2	

Walkerdoo Math Game Puzzle 66 – Medium

Note the pattern in Columns A and B first. Then perform the math operation in each row across and fill in each blank cell

Columns:

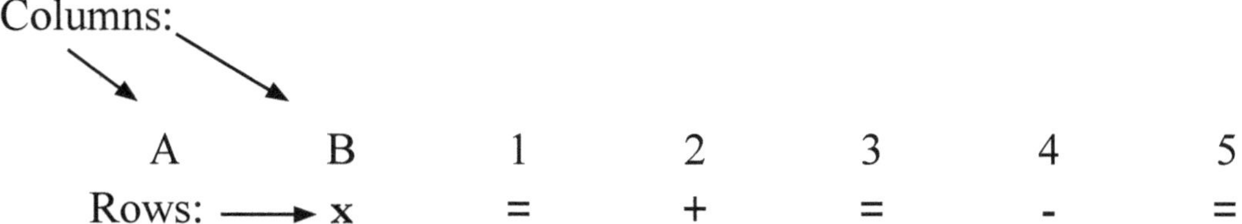

A	B	1	2	3	4	5
x		**=**	**+**	**=**	**-**	**=**

Rows: →

A	B	1	2	3	4	5
8	1		5	13		4
7		7		17	8	
9	1		15		7	
	3	24		44		38
10		30	25		5	
	3	27		57		53
11		55		90		87
	5	50	40		2	
12	5		45	105	1	

Walkerdoo Math Game Puzzle 67 – Medium

Note the pattern in Columns A and B first. Then perform the math operation in each row across and fill in each blank cell

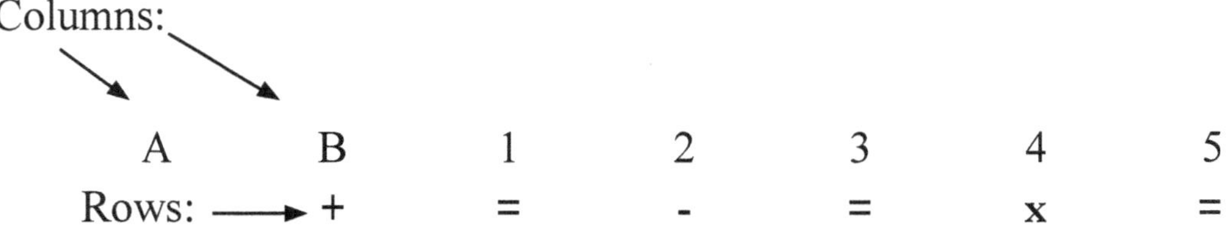

A	B	1	2	3	4	5
5^2	9	34		29	1	
5^2	11		8	28		28
5^2		33	6	27	1	
4^2	10	26		17		34
4^2		23		16	2	
4^2	9	25		15	2	
3^2		15	8		3	
	8	17		6		18
3^2		14		5	3	

Note: To square a number means to multiply the number by itself. For example, 5^2, is 5 x 5 equals 25.

Walkerdoo Math Game Puzzle 68 – Medium

Note the pattern in Columns A and B first. Then perform the math operation in each row across and fill in each blank cell

Columns:

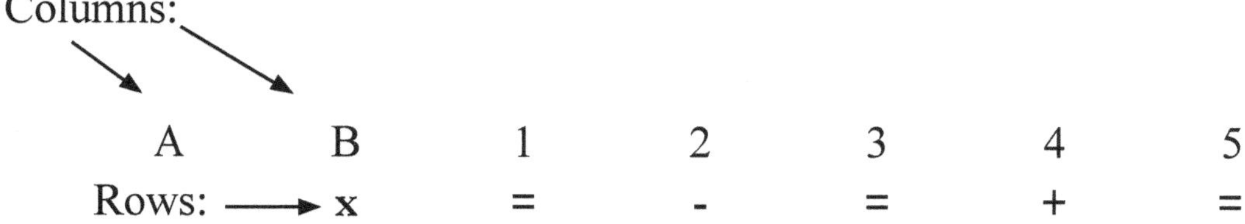

A	B	1	2	3	4	5
x	=	-	=	+	=	

Rows: →

A	B	1	2	3	4	5
12		60		55	5	
8	8		7		10	
10		60		56		71
6	9		6		20	68
	7	56		53	25	
4	10		5	35		65
6		48	2		35	
	11	22		18		58
4		36	1		45	

Walkerdoo Math Game Puzzle 69 – Medium

Note the pattern in Columns A and B first. Then perform the math operation in each row across and fill in each blank cell

Columns:

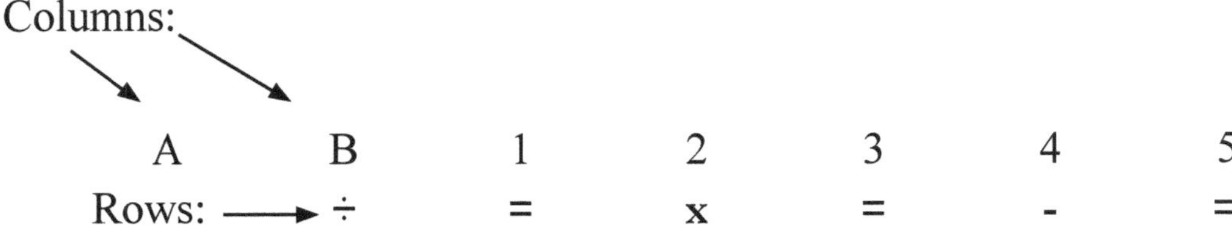

	A	B	1	2	3	4	5
Rows:	÷	=	x	=	-	=	

A ÷	B =	1 x	2 =	3 -	4 =	5
96	8	12		24		19
88		11	5		4	
	8	10		40		33
72		9		63	6	
64	8		6		9	
	8	7		63	8	
48		6		48		37
	8	5	11		10	
32	8		10		13	27

Walkerdoo Math Game Puzzle 70 – Medium

Note the pattern in Columns A and B first. Then perform the math operation in each row across and fill in each blank cell

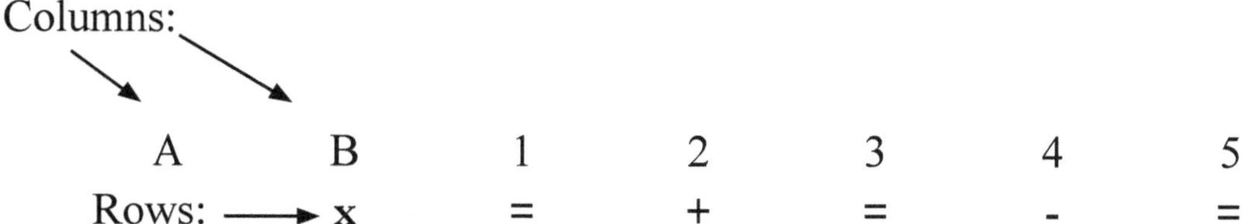

A	B	1	2	3	4	5
3	9	27		30		23
	7	21	6		7	
3	8		4	28		21
	6	18		25	6	
3		21	5		6	
3	5		8	23		17
	6	18		24	5	
3		12	9		5	
	5	15		22		17

Walkerdoo Math Game Puzzle 71 – Medium

Note the pattern in Columns A and B first. Then perform the math
operation in each row across and fill in each blank cell

Columns:

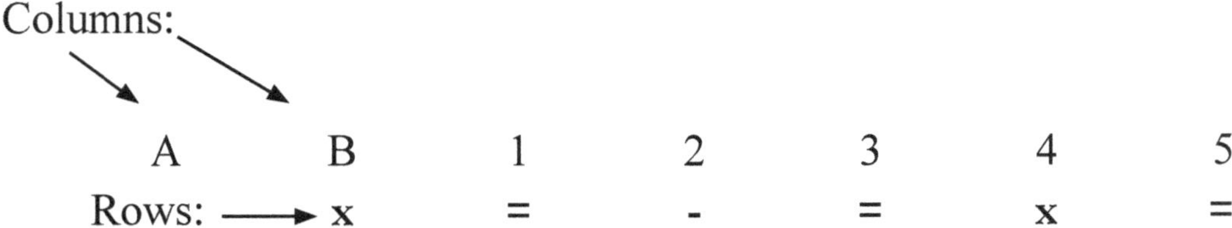

A	B	1	2	3	4	5
x	=	-	=	x	=	

A	B	1	2	3	4	5
12	2	24		12		24
10		30		17	2	
14	2		10		2	36
12		36		25	2	
	2	32		24		48
14		42	9		2	
18		36		30	2	
	3	48		41		82
	2	40		36		72

Walkerdoo Math Game Puzzle 72 – Hard

Note the pattern in Columns A and B first. Then perform the math
operation in each row across and fill in each blank cell

Columns:

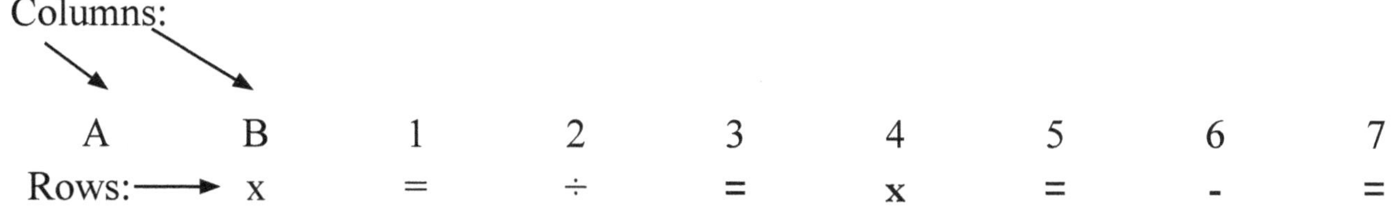

A	B	1	2	3	4	5	6	7
		x	=	÷	=	x	=	-	=

A	B	1	2	3	4	5	6	7
12	1	12	3		9		7	29
9	4		4		3		10	17
10		20		10		70		62
	5	35		7		42	11	
8	3		6	4		20	9	
	6	30		6		24		12
6	4		8		12		10	26
	7	21		7		28	13	
4		20	4		8		11	

Walkerdoo Math Game Puzzle 73 – Hard

Note the pattern in Columns A and B first. Then perform the math operation in each row across and fill in each blank cell

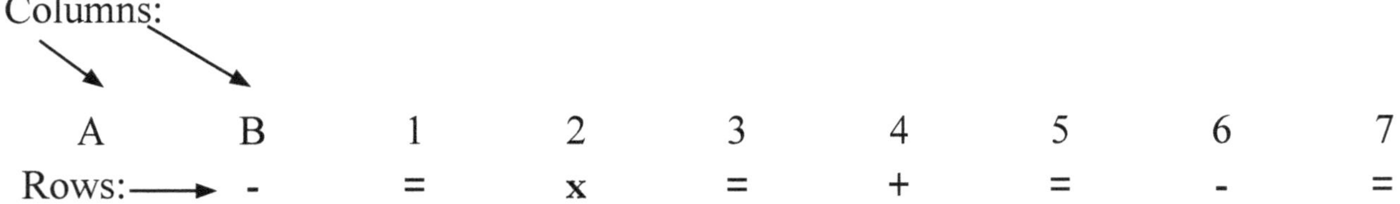

A	B	1	2	3	4	5	6	7
20	5	15		30		36		27
18	7		5		8		10	53
16	4	12		48		58	8	
14		8		56		68		59
	3	9		54		68		61
10		5	9		16		8	
8		6		48		66		60
	4	2		22		42	7	
4		3	10		22		5	

Walkerdoo Math Game Puzzle 74 – Hard

Note the pattern in Columns A and B first. Then perform the math
operation in each row across and fill in each blank cell

Columns:

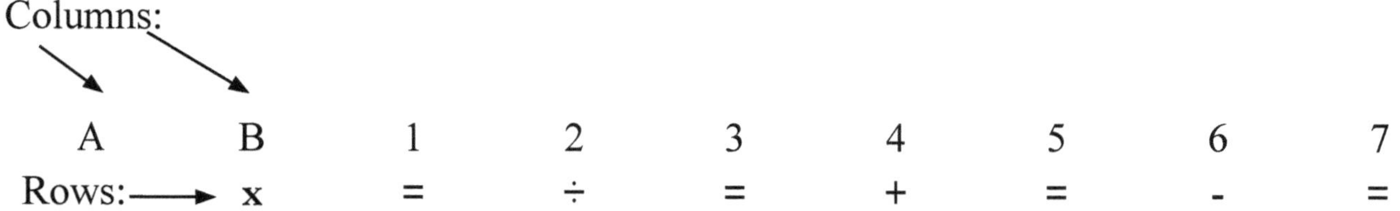

A	B	1	2	3	4	5	6	7
Rows: x	=	÷	=	+	=	-	=	
5	1	5		1		21	12	
	4	20	5		18		11	
5		15	5		16	19		9
	6	36		6		20	9	
6		30	6		12	17		9
6	8		6		10		7	
	7	49		7		15		9
7		70	7		6		5	11
	9	63		9		13	4	

Walkerdoo Math Game Puzzle 75 – Hard

Note the pattern in Columns A and B first. Then perform the math operation in each row across and fill in each blank cell

Columns:

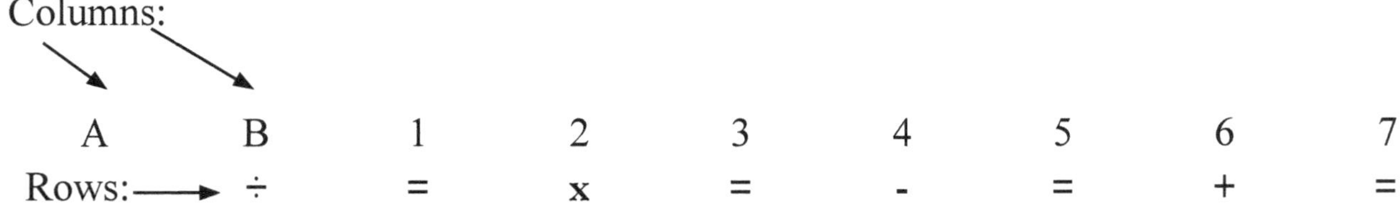

A	B	1	2	3	4	5	6	7
Rows: ÷	=	x	=	-	=	+	=
30	2	15		150		140		142
27	3	9		81		72	4	
24		12	8		8	88		94
21	3		7		7	42	8	
	2	9		54		48	10	
15		5		25		20		32
	2	6		24	4		14	
9		3	3		3		16	
	2	3		6		4		22

Walkerdoo Math Game Puzzle 76 – Hard

Note the pattern in Columns A and B first. Then perform the math operation in each row across and fill in each blank cell

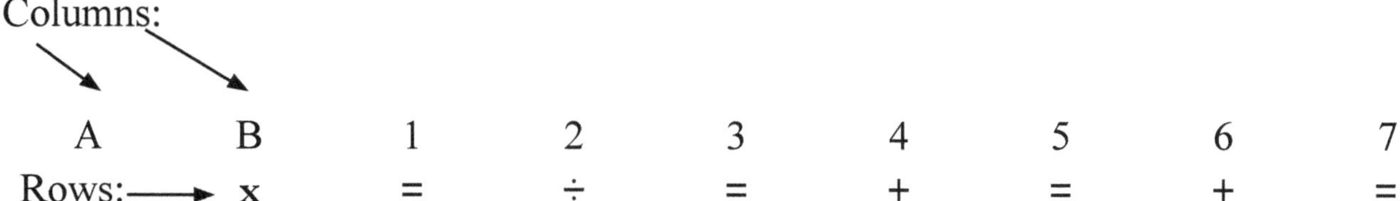

Columns:

A	B	1	2	3	4	5	6	7
Rows: x	=	÷	=	+	=	+	=
5^2	2	50		10		15		26
5^2		75	5		8		10	
5^2	2		5		4	14	9	
4^2		48		12		19		27
4^2	2		4	8		11	7	
4^2		48		12	6		6	
3^2	2		3	6		8		13
3^2		27		9		14	4	
3^2	2	18	3		1	7	3	

Note: To square a number means to multiply the number by itself. For example, 5^2, is 5 x 5 equals 25.

Walkerdoo Math Game Puzzle 77 – Hard

Note the pattern in Columns A and B first. Then perform the math operation in each row across and fill in each blank cell

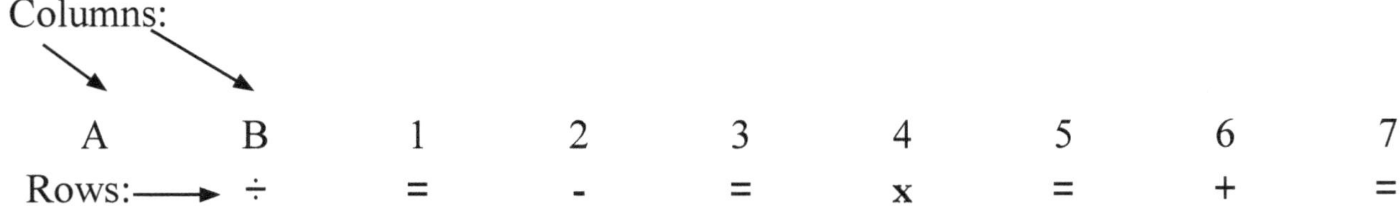

A	B	1	2	3	4	5	6	7
54	2	27		18		36	4	
48		16	8		2	16		21
	2	21		14		28	6	
36		12	6		3		7	25
30	2		5	10		30		38
	3	8		4		12	9	
	2	9	3		4		10	
12		4		2		8		19
	2	3	1		4		12	

Walkerdoo Math Game Puzzle 78 – Hard

Note the pattern in Columns A and B first. Then perform the math operation in each row across and fill in each blank cell

Columns:

A	B	1	2	3	4	5	6	7
(−)		(=)	(x)	(=)	(÷)	(=)	(+)	(=)
9	3	6		54		6	14	
5		2		16		2		15
8		5	7		7		12	
4	2		6		6		11	
	2	5	5		5			15
3		1	4		4		9	
6		5		15		5		13
2	1		2	2		1	7	
5	1		1	4	1		6	10

Rows: →

Walkerdoo Math Game Puzzle 79 – Hard

Note the pattern in Columns A and B first. Then perform the math operation in each row across and fill in each blank cell

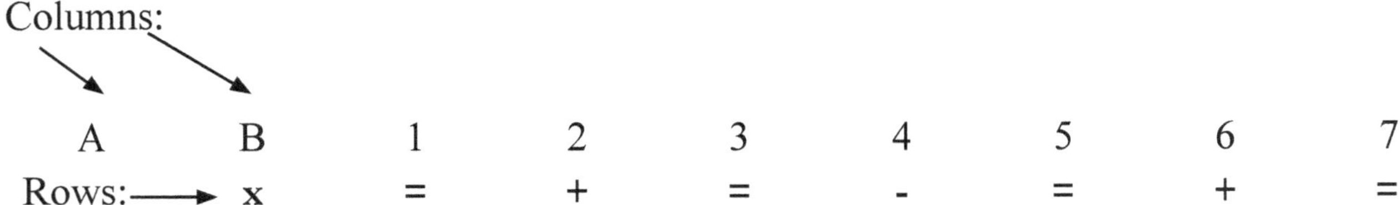

Columns:								
A	B	1	2	3	4	5	6	7
x	=	+	=	-	=	+	=
6	7	42		57	10		3	50
9	5		14	59		50	4	
5		30		43		35		40
8		32	12		7		6	
	5	20		31		25		32
	3	21	10		5		8	
3	4		9	21		17		26
	2	12		20		17	10	
2		6		13	2			22

Walkerdoo Math Game Puzzle 80 – Hard

Note the pattern in Columns A and B first. Then perform the math operation in each row across and fill in each blank cell

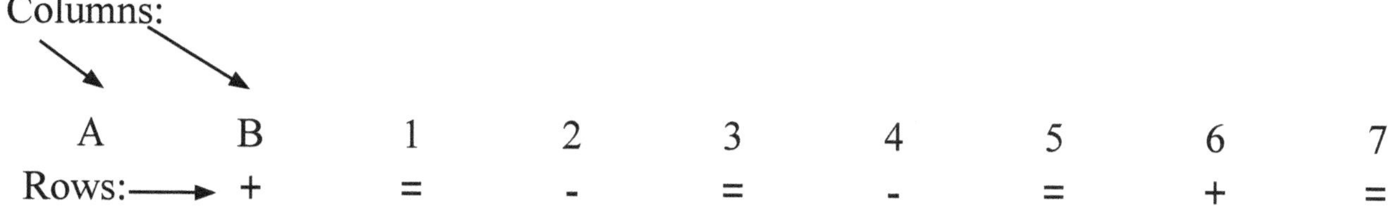

Columns:								
A	B	1	2	3	4	5	6	7
+	=	-	=	-	=	+	=	

A	B	1	2	3	4	5	6	7
9	21	30		28		18		27
	20	25		22		13	8	
8		27	4		8		7	
	18	22		17		10		16
	17	24		18		12		17
3		19	7		5	7	4	
6	15		8		4		3	12
	14	16		7		4	2	
5		18		8	2		1	7

Walkerdoo Math Game Puzzle 81 – Hard

Note the pattern in Columns A and B first. Then perform the math operation in each row across and fill in each blank cell

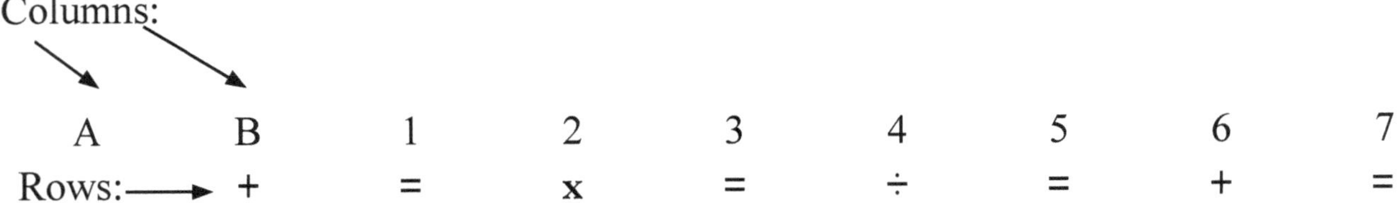

A	B	1	2	3	4	5	6	7
15	9	24		24		6	2^2	10
12		20	1		4	5	2^2	
	7	20		20		5	3^2	14
10		16		32		8	3^2	
	5	16	2		4		4^2	24
8		12		24		6	4^2	22
	3	12		36	4		5^2	
6		8	3		4		5^2	31
	1	8		24		6	5^2	

Note: To square a number means to multiply the number by itself. For example, 2^2, is 2 x 2 equals 4.

Walkerdoo Math Game Puzzle 82 – Challenge

Note the pattern in Columns A and B first. Then perform the math operation in each row across and fill in each blank cell

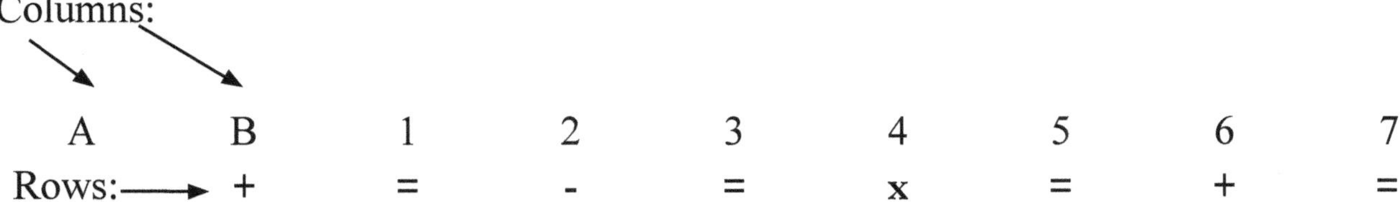

Columns:

A	B	1	2	3	4	5	6	7
+	=	-	=	x	=	+	=	

A	B	1	2	3	4	5	6	7
12	13		10	15		45	2^2	
8		18		9	3		2^2	31
10		21		13		39	2^2	
	8	14	7			28	3^2	
8		17		11	4		3^2	53
	6	10	5			20	3^2	
6		13		9		45	4^2	
	4	6		3	5		4^2	
4		9		7		35	4^2	

Note: To square a number means to multiply the number by itself. For example, 5^2, is 5 x 5 equals 25.

Walkerdoo Math Game Puzzle 83 – Challenge

Note the pattern in Columns A and B first. Then perform the math
operation in each row across and fill in each blank cell

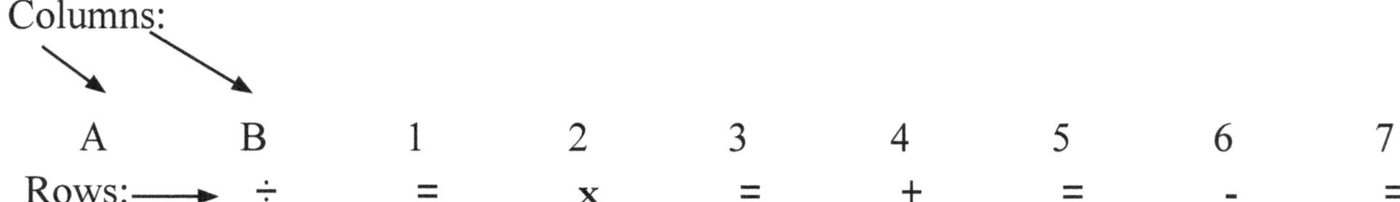

Columns:

A	B	1	2	3	4	5	6	7
÷	=	x	=	+	=	-	=	

A	B	1	2	3	4	5	6	7
77	7	11		22		37		30
	7	10		30		44	5	
63		9	4		13		6	
56		8		40		52		48
	7	7		42		53	5	
42		6	7		10		3	
	7	5		40		49	4	
	7	4		36	8		2	42
21		3		30		37	3	

Motivational Math (Learning Is Fun)

1. Fractions

Proper fraction = is when the numerator is less than the denominator.
Ex: 1/3, 1/8, 2/3, 3/7, 12/13 …..

Improper fraction = is when the numerator is greater than or equal to the denominator. Ex: 6/5, 4/2, 5/3 10/2, 9/9, 11/11, 1000/1000 …. Motivation: Have one student at a time to stand and raise one hand for proper fractions, and two hands for improper fractions. The moderator is to call out as many fractions as possible within twenty seconds. Now try it with 2 to 10 students at a time and select a winner. Also, you may teach adding fractions mentally at this time. Ex: ½ + ½ = 1

2. Buzz – Multiples of 7 and numbers with 7 in it.

Ex: 7, 14, 17, 21, 27, 28, and so on.
Motivation: Choose seven or eight students at a time to come to the head of the class. Have them start counting one number at a time with you (the teacher) as the monitor, pointing at each student. Start with the number one and when they get to multiples of 7 or numbers with seven in it they are to say buzz.
Ex: 1,2,3,4,5,6, buzz 8,9,….
After about 10 rounds of multiples of 7, introduce multiples of 5 with students and have them clap at all multiples of 5, while continuing saying buzz with multiples of 7. Remember, "no one is to clap or say buzz at the same time."

3. 999 with your calculator

Motivation: Take any three-digit number and multiply it by 999.
Ex: 123 X 999 = 122877
Now have one student at a time to extract one number from the answer they have selected, and call the other numbers out one at a time in any order. (1, 7, 2, 2, 7)
You (the teacher) will be able to tell them 8 by adding up all
the numbers called out to you. 1 + 7 + 2 + 2 + 7 = 19
Now take the answer and subtract it from the next multiple of 9.

$$
\begin{array}{r}
\text{EX:} \quad 27 \\
-\ \underline{19} \\
\text{Answer } 8
\end{array}
$$

4. Fun with adding numbers into your calculator

Motivation: Give a calculator to each student. Have them to put into the calculator $1 + 1 = 2$. Make sure each student starts at the same time. Now have students to press the equal sign as fast as they can when you say go! Give students 10 seconds to press the equal sign and see who has the highest number on the calculator afterwards. Increase the minutes each time. The person with the highest number is the winner for that round. Take only reasonable answers.

5. It's your birthday! (Use a calculator)

Have students to put into the calculator their birth month and day that they were born.
For example: Jan 3 = 103, June 26 = 626,
Oct 5 = 1005 or Dec 18 = 1218
Now have students multiply the number by 2, then add 7.
Then they are to multiply the result by 50 and add in their age.
Finally, they are to subtract 3 from the total.
Motivation: Take the calculator from one student at a time and subtract 347 from it. You will then be able to tell them their birthday and age. Ex: 111512, means you were born on the 15th of November and you are 12 years old.

6. Mental Math Relay

Select 10 students and place five on each team. Have students to line up or sit in a straight line and respond within 3 seconds. You (the teacher) will be the announcer and have the students answer the question quickly without raising their hand. Motivation: Two students will stand side by side and whoever responds correctly first, team will receive a point. Both members on different teams are to move to the back of the line and other members are to move forward after each question. This is repeated for many rounds and a team winner is selected. Remember, "both students from each team will move to the back of the line after each response."

7. Fantastic 9

Take any nonrepeating three-digit number and reverse the order of the numbers. Place the larger number at the top and subtract the smaller number. Motivation: You will always have a nine (9) in the middle once you subtract.

Place Value Game

Students will line up in groups of four. Two to four groups at a time will be given a chart of numbers from 0 – 9, with each student having all 10 numbers. The facilitator will call out numbers from one to thousand in standard form. For example, if he called number 2,043, the thousand's family will have a 2, hundreds a 0, tens a 4 and one's a 3. Students will place the called standard form number in order of each family. The first team to have all numbers in order is selected as the winner for that round. The facilitator will decide on how many rounds to be used and identify the overall winner.

Musical Numbers Game

Twelve students will line up to play the musical numbers game. Eleven chairs will be placed around a table, and all twelve students will walk around the table as the facilitator calls out multiplication facts. When the music starts, the game begins. The facilitator will start with any number. For example, he may start out by saying $4 \times 6 = 24$, from that point on the students will continue where the facilitator has stopped for example, ($4 \times 7 = 28$, $4 \times 8 = 32$, etc.) until the music stops. Once the music stops, each student will find a chair to sit down in, and the last one standing will be eliminated. The facilitator will continue this process until the last person is standing.

Walkerdoo Math Puzzles Answer Key– Easy

Note the pattern in Columns A and B. Then perform the math operation in each Row across and fill in each blank cell

1

Columns:	A	B	1	2	3
Rows: →	x	=	+	=
	3	2	6	6	12
	6	2	12	8	20
	4	2	8	5	13
	7	3	21	7	28
	5	3	15	4	19
	8	3	24	6	30
	6	4	24	3	27
	9	4	36	5	41
	7	4	28	2	30

3

Columns:	A	B	1	2	3
Row: →	x	=	-	=
	4	4	16	6	10
	2	2	4	3	1
	5	5	25	5	20
	3	3	9	2	7
	6	6	36	12	24
	4	4	16	5	11
	7	7	49	10	39
	5	5	25	8	17
	8	8	64	9	55

2

Columns:	A	B	1	2	3
Rows: →	x	=	-	=
	6	8	48	7	41
	4	5	20	10	10
	7	7	49	8	41
	5	4	20	11	9
	8	6	48	9	39
	6	3	18	12	6
	9	5	45	10	35
	7	2	14	13	1
	10	4	40	11	29

4

Columns:	A	B	1	2	3
Rows: →	+	=	-	=
	1^2	5	6	2	4
	2^2	8	12	6	6
	3^2	6	15	7	8
	4^2	9	25	20	5
	5^2	7	32	7	25
	6^2	10	46	16	30
	7^2	8	57	7	50
	8^2	11	75	10	65
	9^2	9	90	15	75

Note: To square a number means to multiply the number by itself. For example, 5^2, is 5 x 5 equals 25.

Walkerdoo Math Puzzles Answer Key– Easy

Note the pattern or sequence in Columns A or B. Then perform the
math operation in each Row across and fill in each blank

Puzzle 5

Columns:	A	B	1	2	3
Rows: →		+	=	x	=
	2	9	11	3	33
	5	8	13	4	52
	3	7	10	5	50
	6	6	12	6	72
	4	5	9	7	63
	7	4	11	8	88
	5	3	8	9	72
	8	2	10	10	100
	6	1	7	11	77

Puzzle 7

Columns:	A	B	1	2	3
Row: →		x	=	÷	=
	10	3	30	3	10
	9	3	27	3	9
	8	3	24	3	8
	7	3	21	3	7
	6	3	18	3	6
	5	3	15	3	5
	4	3	12	3	4
	3	3	9	3	3
	2	3	6	3	2

Puzzle 6

Columns:	A	B	1	2	3
Rows: →		x	=	-	=
	1	9	9	5	4
	4	8	32	4	28
	2	7	14	3	11
	5	6	30	2	28
	3	5	15	1	14
	6	4	24	2	22
	4	3	12	3	9
	7	2	14	4	10
	5	1	5	5	0

Puzzle 8

Columns:	A	B	1	2	3
Rows: →		x	=	+	=
	2	1	2	9	11
	3	2	6	8	14
	4	3	12	7	19
	5	4	20	6	26
	6	5	30	5	35
	7	6	42	4	46
	8	7	56	3	59
	9	8	72	2	74
	10	9	90	1	91

Walkerdoo Math Puzzles Answer Key– Easy

Note the pattern or sequence in Columns A or B. Then perform the math operation in each Row across and fill in each blank

Puzzle 9 — Columns: A, B, 1, 2, 3; Rows: → x = ÷ =

A	B	1	2	3
6	11	66	6	11
6	10	60	6	10
6	9	54	6	9
6	8	48	6	8
6	7	42	6	7
6	6	36	6	6
6	5	30	6	5
6	4	24	6	4
6	3	18	6	3

Puzzle 11 — Columns: A, B, 1, 2, 3; Row: → x = + =

A	B	1	2	3
6	1	6	14	20
3	2	6	13	19
5	3	15	12	27
2	4	8	11	19
4	5	20	10	30
1	6	6	9	15
3	7	21	8	29
0	8	0	7	7
2	9	18	6	24

Puzzle 10 — Columns: A, B, 1, 2, 3; Rows: → + = + =

A	B	1	2	3
10	1	11	9	20
9	3	12	8	20
8	2	10	7	17
7	4	11	6	17
6	3	9	5	14
5	5	10	4	14
4	4	8	3	11
3	6	9	2	11
2	5	7	1	8

Puzzle 12 — Columns: A, B, 1, 2, 3; Rows: → x = - =

A	B	1	2	3
8	9	72	9	63
5	8	40	8	32
7	7	49	7	42
4	6	24	6	18
6	5	30	5	25
3	4	12	4	8
5	3	15	3	12
2	2	4	2	2
4	1	4	1	3

Walkerdoo Math Puzzles Answer Key– Easy

Note the pattern or sequence in Columns A or B. Then perform the
math operation in each Row across and fill in each blank

Puzzle 13

Columns:	A	B	1	2	3
Rows: →		+	=	+	=
	18	12	30	2	32
	16	11	27	4	31
	14	10	24	6	30
	12	9	21	8	29
	10	8	18	10	28
	8	7	15	12	27
	6	6	12	14	26
	4	5	9	16	25
	2	4	6	18	24

Puzzle 15

Columns:	A	B	1	2	3
Row: →		÷	=	x	=
	81	9	9	2	18
	72	9	8	5	40
	63	9	7	4	28
	54	9	6	7	42
	45	9	5	6	30
	36	9	4	9	36
	27	9	3	8	24
	18	9	2	11	22
	9	9	1	10	10

Puzzle 14

Columns:	A	B	1	2	3
Rows: →		÷	=	x	=
	12	6	2	12	24
	10	5	2	11	22
	14	7	2	10	20
	12	6	2	9	18
	16	8	2	8	16
	14	7	2	7	14
	18	9	2	6	12
	16	8	2	5	10
	20	10	2	4	8

Puzzle 16

Columns:	A	B	1	2	3
Rows: →		+	=	x	=
	9^2	9	90	1	90
	6^2	8	44	2	88
	8^2	7	71	1	71
	5^2	6	31	2	62
	7^2	5	54	1	54
	4^2	4	20	2	40
	6^2	3	39	1	39
	3^2	2	11	2	22
	5^2	1	26	1	26

Note: To square a number means to multiply the number by itself. For example, 5^2, is 5 x 5 equals 25.

Walkerdoo Math Puzzles Answer Key- Easy

Note the pattern or sequence in Columns A or B. Then perform the
math operation in each Row across and fill in each blank

Puzzle 17

Columns:	A	B	1	2	3
Rows: →		x	=	÷	=
	4	11	44	4	11
	4	10	40	4	10
	4	9	36	4	9
	4	8	32	4	8
	4	7	28	4	7
	4	6	24	4	6
	4	5	20	4	5
	4	4	16	4	4
	4	3	12	4	3

Puzzle 19

Columns:	A	B	1	2	3
Row: →		+	=	x	=
	1	5	6	8	48
	2	8	10	5	50
	3	4	7	7	49
	4	7	11	4	44
	5	3	8	6	48
	6	6	12	3	36
	7	2	9	5	45
	8	5	13	2	26
	9	1	10	4	40

Puzzle 18

Columns:	A	B	1	2	3
Rows: →		x	=	+	=
	1	9	9	4	13
	4	8	32	2	34
	2	7	14	6	20
	5	6	30	4	34
	3	5	15	8	23
	6	4	24	6	30
	4	3	12	10	22
	7	2	14	8	22
	5	1	5	12	17

Puzzle 20

Columns:	A	B	1	2	3
Rows: →		÷	=	x	=
	30	2	15	3	45
	27	3	9	3	27
	24	2	12	3	36
	21	3	7	4	28
	18	2	9	4	36
	15	3	5	4	20
	12	2	6	5	30
	9	3	3	5	15
	6	2	3	5	15

Walkerdoo Math Puzzles Answer Key– Easy

Note the pattern or sequence in Columns A or B. Then perform the math operation in each Row across and fill in each blank

Puzzle 21

Columns:	A	B	1	2	3
Rows: →		x	=	-	=
	2	9	18	9	9
	2	8	16	9	7
	3	7	21	7	14
	3	6	18	7	11
	4	5	20	5	15
	4	4	16	5	11
	5	3	15	3	12
	5	2	10	3	7
	6	1	6	1	5

Puzzle 23

Columns:	A	B	1	2	3
Row: →		+	=	-	=
	4	5	9	3	6
	4	1	5	3	2
	5	7	12	3	9
	5	3	8	4	4
	6	9	15	4	11
	6	5	11	4	7
	7	11	18	5	13
	7	7	14	5	9
	7	13	20	5	15

Puzzle 22

Columns:	A	B	1	2	3
Rows: →		+	=	-	=
	3	15	18	2	16
	3	12	15	3	12
	4	13	17	2	15
	4	10	14	3	11
	5	11	16	2	14
	5	8	13	3	10
	6	9	15	2	13
	6	6	12	3	9
	7	7	14	2	12

Puzzle 24

Columns:	A	B	1	2	3
Rows: →		+	=	-	=
	1^3	9	10	6	4
	1^3	8	9	6	3
	1^3	7	8	6	2
	2^3	6	14	8	6
	2^3	5	13	8	5
	2^3	4	12	8	4
	3^3	3	30	10	20
	3^3	2	29	10	19
	3^3	1	28	10	18

Note: To cube a number means to multiply the number by itself three times. For example, 3^3 or 3 cubed means, 3 x 3 x 3 = 27.

Walkerdoo Math Puzzles Answer Key– Easy

Note the pattern or sequence in Columns A or B. Then perform the
math operation in each Row across and fill in each blank

Puzzle 25

Columns:	A	B	1	2	3
Rows: →		x	=	+	=
	5	1	5	15	20
	7	2	14	14	28
	4	3	12	13	25
	6	4	24	12	36
	3	5	15	11	26
	5	6	30	10	40
	2	7	14	9	23
	4	8	32	8	40
	1	9	9	7	16

Puzzle 27

Columns:	A	B	1	2	3
Row: →		x	=	÷	=
	12	6	72	6	12
	11	6	66	6	11
	10	6	60	6	10
	9	6	54	6	9
	8	6	48	6	8
	7	6	42	6	7
	6	6	36	6	6
	5	6	30	6	5
	4	6	24	6	4

Puzzle 26

Columns:	A	B	1	2	3
Rows: →		x	=	+	=
	1	9	9	12	21
	3	8	24	11	35
	2	7	14	10	24
	4	6	24	9	33
	3	5	15	8	23
	5	4	20	7	27
	4	3	12	6	18
	6	2	12	5	17
	5	1	5	4	9

Puzzle 28

Columns:	A	B	1	2	3
Rows: →		+	=	+	=
	4	1	5	18	23
	4	4	8	20	28
	4	3	7	16	23
	5	6	11	18	29
	5	5	10	14	24
	5	8	13	16	29
	6	7	13	12	25
	6	10	16	14	30
	6	9	15	10	25

Walkerdoo Math Puzzles Answer Key– Easy

Note the pattern or sequence in Columns A or B. Then perform the math operation in each Row across and fill in each blank

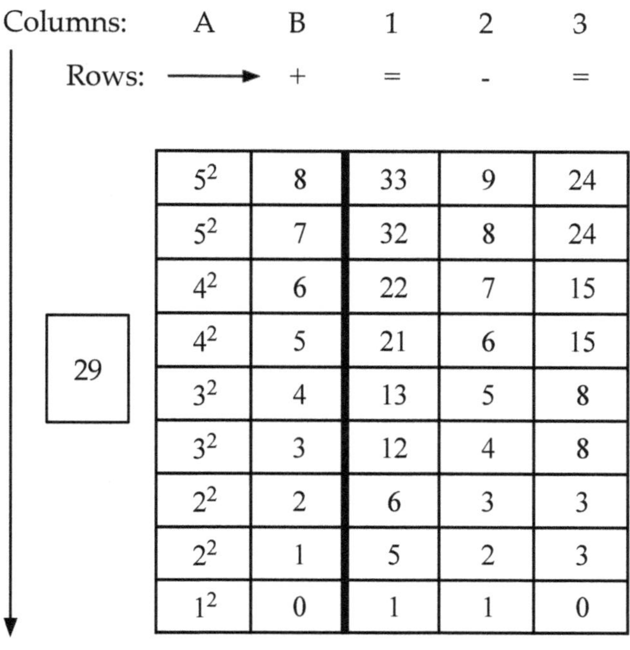

29

Columns:	A	B	1	2	3
Rows: →		+	=	-	=
	5^2	8	33	9	24
	5^2	7	32	8	24
	4^2	6	22	7	15
	4^2	5	21	6	15
	3^2	4	13	5	8
	3^2	3	12	4	8
	2^2	2	6	3	3
	2^2	1	5	2	3
	1^2	0	1	1	0

31

Columns:	A	B	1	2	3
Row: →		÷	=	x	=
	30	2	15	10	150
	27	3	9	9	81
	24	2	12	8	96
	21	3	7	7	49
	18	2	9	6	54
	15	3	5	5	25
	12	2	6	4	24
	9	3	3	3	9
	6	2	3	2	6

Note: To square a number means to multiply the number times itself. For example, 5^2, is 5 x 5 equals 25.

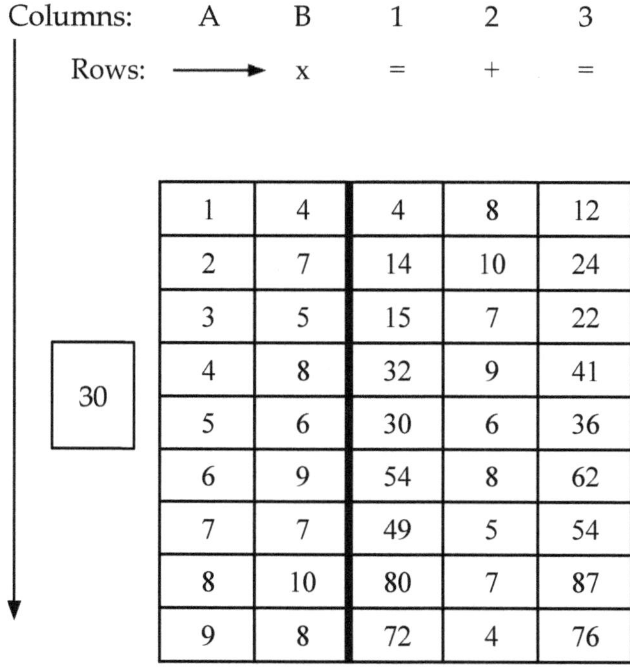

30

Columns:	A	B	1	2	3
Rows: →		x	=	+	=
	1	4	4	8	12
	2	7	14	10	24
	3	5	15	7	22
	4	8	32	9	41
	5	6	30	6	36
	6	9	54	8	62
	7	7	49	5	54
	8	10	80	7	87
	9	8	72	4	76

32

Columns:	A	B	1	2	3
Rows: →		+	=	x	=
	9	0	9	4	36
	7	1	8	4	32
	10	2	12	4	48
	8	3	11	3	33
	11	4	15	3	45
	9	5	14	3	42
	12	6	18	2	36
	10	7	17	2	34
	13	8	21	2	42

Walkerdoo Math Puzzles Answer Key– Easy

Note the pattern or sequence in Columns A or B. Then perform the
math operation in each Row across and fill in each blank

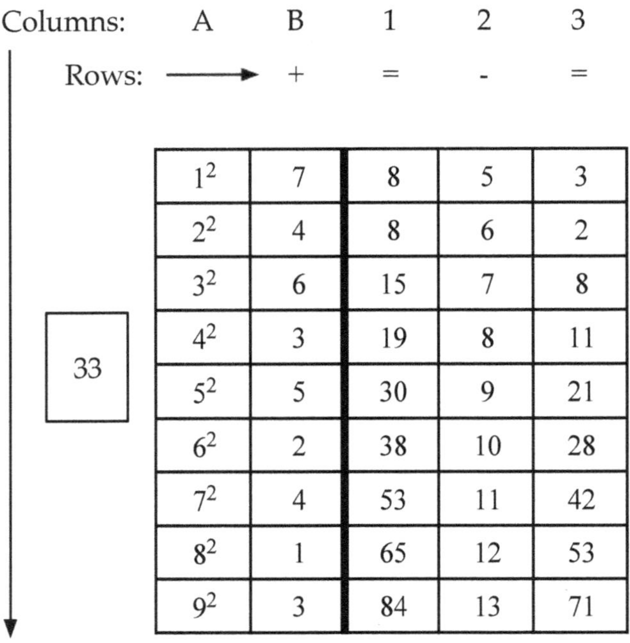

Puzzle 33

Columns:	A	B	1	2	3
Rows: →	+	=	-	=	

A	B	1	2	3
1^2	7	8	5	3
2^2	4	8	6	2
3^2	6	15	7	8
4^2	3	19	8	11
5^2	5	30	9	21
6^2	2	38	10	28
7^2	4	53	11	42
8^2	1	65	12	53
9^2	3	84	13	71

Note: To square a number means to multiply the number by itself. For example, 5^2, is 5 x 5 equals 25.

Puzzle 35

Columns:	A	B	1	2	3
Row: →	x	=	÷	=	

A	B	1	2	3
6	6	36	6	6
8	8	64	8	8
5	5	25	5	5
7	7	49	7	7
4	4	16	4	4
6	6	36	6	6
3	3	9	3	3
5	5	25	5	5
2	2	4	2	2

Puzzle 34

Columns:	A	B	1	2	3
Rows: →	x	=	÷	=	

A	B	1	2	3
2	9	18	2	9
3	8	24	3	8
4	7	28	2	14
5	6	30	3	10
6	5	30	2	15
7	4	28	2	14
8	3	24	3	8
9	2	18	3	6
10	1	10	2	5

Puzzle 36

Columns:	A	B	1	2	3
Rows: →	+	=	-	=	

A	B	1	2	3
9	2^2	13	7	6
8	2^2	12	7	5
7	3^2	16	7	9
6	3^2	15	5	10
5	4^2	21	5	16
4	4^2	20	5	15
3	5^2	28	3	25
2	5^2	27	3	24
1	6^2	37	3	34

Note: To square a number means to multiply the number by itself. For example, 5^2, is 5 x 5 equals 25.

Walkerdoo Math Puzzles Answer Key– Easy

Note the pattern or sequence in Columns A or B. Then perform the
math operation in each Row across and fill in each blank

Puzzle 37

Columns:	A	B	1	2	3
Rows: →		x	=	÷	=
	12	7	84	7	12
	11	7	77	7	11
	10	7	70	7	10
	9	7	63	7	9
	8	7	56	7	8
	7	7	49	7	7
	6	7	42	7	6
	5	7	35	7	5
	4	7	28	7	4

Puzzle 39

Columns:	A	B	1	2	3
Row: →		+	=	-	=
	10	2^3	18	9	9
	12	2^3	20	8	12
	9	2^3	17	7	10
	11	3^3	38	6	32
	8	3^3	35	5	30
	10	3^3	37	4	33
	7	4^3	71	3	68
	9	4^3	73	2	71
	6	4^3	70	1	69

Puzzle 38

Columns:	A	B	1	2	3
Rows: →		-	=	x	=
	7	2	5	9	45
	8	4	4	8	32
	9	3	6	7	42
	10	5	5	6	30
	11	4	7	5	35
	12	6	6	4	24
	13	5	8	3	24
	14	7	7	2	14
	15	6	9	1	9

Note: To cube a number means to multiply
the number by itself three times. For example, 3^3
or 3 cubed means, 3 x 3 x 3 = 27.

Puzzle 40

Columns:	A	B	1	2	3
Rows: →		x	=	+	=
	1	2	2	12	14
	2	2	4	8	12
	3	2	6	10	16
	4	3	12	6	18
	5	3	15	8	23
	6	3	18	4	22
	7	4	28	6	34
	8	4	32	2	34
	9	4	36	4	40

Walkerdoo Math Puzzles Answer Key– Easy

Note the pattern or sequence in Columns A or B. Then perform the math operation in each Row across and fill in each blank

Puzzle 41

Columns:	A	B	1	2	3
Rows: →		+	=	x	=
	9	15	24	2	48
	8	12	20	3	60
	7	13	20	2	40
	6	10	16	3	48
	5	11	16	2	32
	4	8	12	3	36
	3	9	12	2	24
	2	6	8	3	24
	1	7	8	2	16

Puzzle 43

Columns:	A	B	1	2	3
Row: →		x	=	+	=
	5	1	5	12	17
	5	4	20	8	28
	5	3	15	10	25
	6	6	36	6	42
	6	5	30	8	38
	6	8	48	4	52
	7	7	49	6	55
	7	10	70	2	72
	7	9	63	4	67

Puzzle 42

Columns:	A	B	1	2	3
Rows: →		÷	=	x	=
	80	8	10	5	50
	72	8	9	4	36
	64	8	8	7	56
	56	8	7	6	42
	48	8	6	9	54
	40	8	5	8	40
	32	8	4	11	44
	24	8	3	10	30
	16	8	2	13	26

Puzzle 44

Columns:	A	B	1	2	3
Rows: →		+	=	-	=
	5^2	5	30	15	15
	5^2	8	33	6	27
	5^2	6	31	7	24
	4^2	9	25	20	5
	4^2	7	23	7	16
	4^2	10	26	16	10
	3^2	8	17	8	9
	3^2	11	20	10	10
	3^2	9	18	3	15

Note: To square a number means to multiply the by times itself. For example, 5^2, is 5 x 5 equals 25.

Walkerdoo Math Puzzles Answer Key– Easy

Follow the pattern or sequence in Columns A or B. Then perform
the arithmetic operation in each Row and fill in each blank

Puzzle 45

Columns:	A	B	1	2	3
Rows: →		x	=	-	=
	9	5	45	9	36
	8	8	64	8	56
	7	4	28	7	21
	6	7	42	6	36
	5	3	15	5	10
	4	6	24	4	20
	3	2	6	3	3
	2	5	10	2	8
	1	1	1	1	0

Puzzle 47

Columns:	A	B	1	2	3
Row: →		÷	=	-	=
	36	2	18	4	14
	33	3	11	4	7
	30	2	15	4	11
	27	3	9	3	6
	24	2	12	3	9
	21	3	7	3	4
	18	2	9	2	7
	15	3	5	2	3
	12	2	6	2	4

Puzzle 46

Columns:	A	B	1	2	3
Rows: →		+	=	-	=
	9^2	2	83	9	74
	6^2	3	39	8	31
	8^2	4	68	7	61
	5^2	5	30	6	24
	7^2	6	55	5	50
	4^2	7	23	4	19
	6^2	8	44	3	41
	3^2	9	18	2	16
	5^2	10	35	1	34

Puzzle 48

Columns:	A	B	1	2	3
Rows: →		x	=	-	=
	8	5	40	9	31
	6	8	48	8	40
	9	6	54	7	47
	7	9	63	6	57
	10	7	70	5	65
	8	10	80	4	76
	11	8	88	3	85
	9	11	99	2	97
	12	9	108	1	107

Note: To square a number means to
multiply the number by itself. For example,
5^2, is 5 x 5 equals 25.

Walkerdoo Math Puzzles Answer Key- Easy

Note the pattern or sequence in Columns A or B. Then perform the
math operation in each Row across and fill in each blank

Puzzle 49

Columns:	A	B	1	2	3
Rows: →	+	=	-	=	
	2^3	12	20	9	11
	2^3	14	22	8	14
	2^3	11	19	7	12
	3^3	9	36	6	30
	3^3	6	33	5	28
	3^3	8	35	4	31
	4^3	5	69	3	66
	4^3	7	71	2	69
	4^3	4	68	1	67

Puzzle 51

Columns:	A	B	1	2	3
Row: →	+	=	x	=	
	12	1	13	2	26
	8	4	12	2	24
	10	2	12	3	36
	6	5	11	3	33
	8	3	11	4	44
	4	6	10	4	40
	6	4	10	5	50
	2	7	9	5	45
	4	5	9	6	54

Note: To cube a number means to multiply
the number by itself three times. For example,
3^3 or 3 cubed means, 3 x 3 x 3 = 27.

Puzzle 50

Columns:	A	B	1	2	3
Rows: →	-	=	x	=	
	6	2	4	9	36
	7	4	3	10	30
	8	3	5	7	35
	9	5	4	8	32
	10	4	6	5	30
	11	6	5	6	30
	12	5	7	3	21
	13	7	6	4	24
	14	6	8	1	8

Puzzle 52

Columns:	A	B	1	2	3
Rows: →	÷	=	+	=	
	42	2	21	9	30
	39	3	13	8	21
	36	2	18	7	25
	33	3	11	6	17
	30	2	15	5	20
	27	3	9	4	13
	24	2	12	3	15
	21	3	7	2	9
	18	2	9	1	10

Walkerdoo Math Puzzles Answer Key– Easy

Note the pattern or sequence in Columns A or B. Then perform the
math operation in each Row across and fill in each blank

53

Columns:	A	B	1	2	3
Rows: →		x	=	-	=
	2	9	18	11	7
	5	8	40	10	30
	3	7	21	9	12
	6	6	36	8	28
	4	5	20	7	13
	7	4	28	6	22
	5	3	15	5	10
	8	2	16	4	12
	6	1	6	3	3

55

Columns:	A	B	1	2	3
Row: →		x	=	÷	=
	5	4	20	4	5
	9	4	36	4	9
	6	4	24	4	6
	10	5	50	5	10
	7	5	35	5	7
	11	5	55	5	11
	8	6	48	6	8
	12	6	72	6	12
	9	6	54	6	9

54

Columns:	A	B	1	2	3
Rows: →		+	=	÷	=
	81	9	90	9	10
	72	9	81	9	9
	63	9	72	9	8
	54	9	63	9	7
	45	9	54	9	6
	36	9	45	9	5
	27	9	36	9	4
	18	9	27	9	3
	9	9	18	9	2

Walkerdoo Math Puzzles Answer Key – Medium

Note the pattern in Columns A and B first. Then perform the math operation in each row across and fill in each blank cell

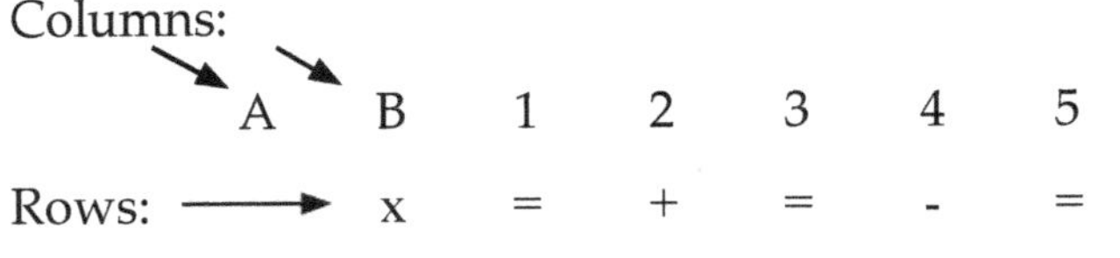

Columns: (left) A B 1 2 3 4 5
Rows: →   x = + = - =

Columns: (right) A B 1 2 3 4 5
Row: →   + = x = - =

56

A	B	1	2	3	4	5
1	8	8	5	13	9	4
2	7	14	10	24	8	16
3	6	18	15	33	7	26
4	5	20	20	40	6	34
5	4	20	25	45	5	40
6	3	18	30	48	4	44
7	2	14	35	49	3	46
8	1	8	40	48	2	46
9	0	0	45	45	1	44

58

A	B	1	2	3	4	5
10	1	11	9	99	10	89
9	3	12	8	96	9	87
8	2	10	7	70	8	62
7	4	11	6	66	7	59
6	3	9	5	45	6	39
5	5	10	4	40	5	35
4	4	8	3	24	4	20
3	6	9	2	18	3	15
2	5	7	1	7	2	5

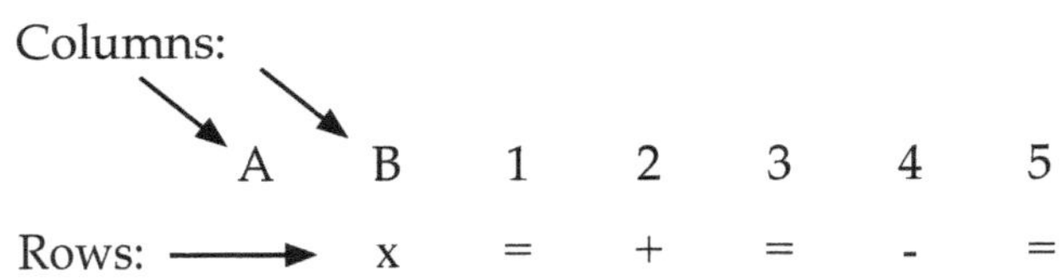

Columns: (left) A B 1 2 3 4 5
Rows: →   x = + = - =

Columns: (right) A B 1 2 3 4 5
Rows: →   ÷ = x = - =

57

A	B	1	2	3	4	5
2	9	18	5	23	10	13
5	7	35	10	45	9	36
3	8	24	15	39	8	31
6	6	36	20	56	7	49
4	7	28	25	53	6	47
7	5	35	30	65	5	60
5	6	30	35	65	4	61
8	4	32	40	72	3	69
6	5	30	45	75	2	73

59

A	B	1	2	3	4	5
99	11	9	2	18	8	10
88	11	8	5	40	5	35
77	11	7	4	28	7	21
66	11	6	7	42	4	38
55	11	5	6	30	6	24
44	11	4	9	36	3	33
33	11	3	8	24	5	19
22	11	2	11	22	2	20
11	11	1	10	10	4	6

Walkerdoo Math Puzzles Answer Key– Medium

Note the pattern in Columns A and B first. Then perform the math operation in each row across and fill in each blank cell

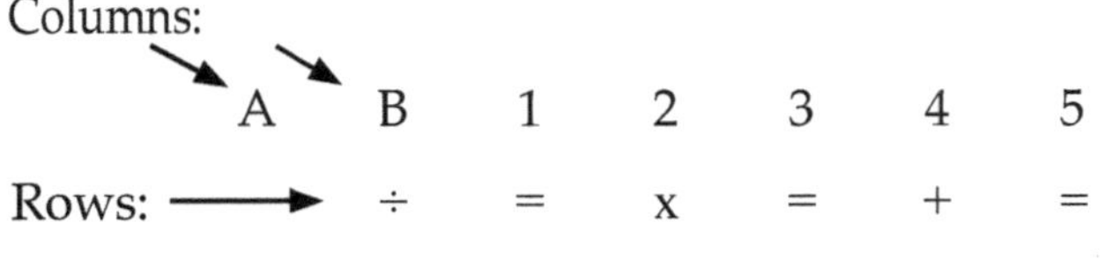

Columns: A → B

Rows: → ÷ = x = + =

60

A	B	1	2	3	4	5
9^2	9	9	3	27	8	35
6^2	6	6	4	24	5	29
8^2	8	8	5	40	7	47
5^2	5	5	6	30	4	34
7^2	7	7	7	49	6	55
4^2	4	4	8	32	3	35
6^2	6	6	9	54	5	59
3^2	3	3	10	30	2	32
5^2	5	5	11	55	4	59

Note: To square a number means to multiply the number by itself. For example, 9^2, is 9 x 9 equals 81.

Columns: A → B

Row: → ÷ = + = - =

62

A	B	1	2	3	4	5
36	2	18	12	30	3	27
33	3	11	13	24	3	21
30	2	15	10	25	3	22
27	3	9	11	20	4	16
24	2	12	8	20	4	16
21	3	7	9	16	4	12
18	2	9	6	15	5	10
15	3	5	7	12	5	7
12	2	6	4	10	5	5

Columns: A → B

Rows: → x = x = ÷ =

61

A	B	1	2	3	4	5
6	8	48	2	96	2	48
9	5	45	2	90	2	45
5	7	35	2	70	2	35
8	4	32	3	96	3	32
4	6	24	3	72	3	24
7	3	21	3	63	3	21
3	5	15	4	60	4	15
6	2	12	4	48	4	12
2	4	8	4	32	4	8

Columns: A → B

Rows: → + = + = - =

63

A	B	1	2	3	4	5
2^3	9	17	12	29	10	19
2^3	8	16	13	29	9	20
2^3	7	15	10	25	8	17
3^3	6	33	11	44	7	37
3^3	5	32	8	40	6	34
3^3	4	31	9	40	5	35
4^3	3	67	6	73	4	69
4^3	2	66	7	73	3	70
4^3	1	65	4	69	2	67

Note: To cube a number means to multiply the number by itself three times. For example, 3^3 or 3 cubed means, 3 x 3 x 3 = 27.

Walkerdoo Math Puzzles Answer Key- Medium

Note the pattern in Columns A and B first. Then perform the math operation in each row across and fill in each blank cell

Columns: A → B

64 — Rows: x = + = - =

A	B	1	2	3	4	5
4	1	4	10	14	4	10
4	4	16	16	32	7	25
4	3	12	14	26	5	21
5	6	30	20	50	8	42
5	5	25	18	43	6	37
5	8	40	24	64	9	55
6	7	42	22	64	7	57
6	10	60	28	88	10	78
6	9	54	26	80	8	72

Columns: A → B

66 — Row: x = + = - =

A	B	1	2	3	4	5
8	1	8	5	13	9	4
7	1	7	10	17	8	9
9	1	9	15	24	7	17
8	3	24	20	44	6	38
10	3	30	25	55	5	50
9	3	27	30	57	4	53
11	5	55	35	90	3	87
10	5	50	40	90	2	88
12	5	60	45	105	1	104

Columns: A → B

65 — Rows: + = + = x =

A	B	1	2	3	4	5
4	11	15	6	21	2	42
7	10	17	7	24	2	48
5	9	14	8	22	2	44
8	8	16	9	25	2	50
6	7	13	10	23	2	46
9	6	15	11	26	2	52
7	5	12	12	24	2	48
10	4	14	13	27	2	54
8	3	11	14	25	2	50

Columns: A → B

67 — Rows: + = - = x =

A	B	1	2	3	4	5
5^2	9	34	5	29	1	29
5^2	11	36	8	28	1	28
5^2	8	33	6	27	1	27
4^2	10	26	9	17	2	34
4^2	7	23	7	16	2	32
4^2	9	25	10	15	2	30
3^2	6	15	8	7	3	21
3^2	8	17	11	6	3	18
3^2	5	14	9	5	3	15

Note: To square a number means to multiply the number by itself. For example, 5^2, is 5 x 5 equals 25.

Walkerdoo Math Puzzles Answer Key– Medium

Note the pattern in Columns A and B first. Then perform the math operation in each row across and fill in each blank cell

Puzzle 68

Columns: A B 1 2 3 4 5
Rows: x = - = + =

A	B	1	2	3	4	5
12	5	60	5	55	5	60
8	8	64	7	57	10	67
10	6	60	4	56	15	71
6	9	54	6	48	20	68
8	7	56	3	53	25	78
4	10	40	5	35	30	65
6	8	48	2	46	35	81
2	11	22	4	18	40	58
4	9	36	1	35	45	80

Puzzle 70

Columns: A B 1 2 3 4 5
Row: x = + = - =

A	B	1	2	3	4	5
3	9	27	3	30	7	23
3	7	21	6	27	7	20
3	8	24	4	28	7	21
3	6	18	7	25	6	19
3	7	21	5	26	6	20
3	5	15	8	23	6	17
3	6	18	6	24	5	19
3	4	12	9	21	5	16
3	5	15	7	22	5	17

Puzzle 69

Columns: A B 1 2 3 4 5
Rows: ÷ = x = - =

A	B	1	2	3	4	5
96	8	12	2	24	5	19
88	8	11	5	55	4	51
80	8	10	4	40	7	33
72	8	9	7	63	6	57
64	8	8	6	48	9	39
56	8	7	9	63	8	55
48	8	6	8	48	11	37
40	8	5	11	55	10	45
32	8	4	10	40	13	27

Puzzle 71

Columns: A B 1 2 3 4 5
Rows: x = - = x =

A	B	1	2	3	4	5
12	2	24	12	12	2	24
10	3	30	13	17	2	34
14	2	28	10	18	2	36
12	3	36	11	25	2	50
16	2	32	8	24	2	48
14	3	42	9	33	2	66
18	2	36	6	30	2	60
16	3	48	7	41	2	82
20	2	40	4	36	2	72

Walkerdoo Math Puzzles Answer Key – Hard

Note the pattern in Columns A and B first. Then perform the math operation in each row across and fill in each blank cell

72 — Columns A, B; Rows: x = ÷ = x = - =

A	B	1	2	3	4	5	6	7
12	1	12	3	4	9	36	7	29
9	4	36	4	9	3	27	10	17
10	2	20	2	10	7	70	8	62
7	5	35	5	7	6	42	11	31
8	3	24	6	4	5	20	9	11
5	6	30	5	6	4	24	12	12
6	4	24	8	3	12	36	10	26
3	7	21	3	7	4	28	13	15
4	5	20	4	5	8	40	11	29

74 — Columns A, B; Row: x = ÷ = + = - =

A	B	1	2	3	4	5	6	7
5	1	5	5	1	20	21	12	9
5	4	20	5	4	18	22	11	11
5	3	15	5	3	16	19	10	9
6	6	36	6	6	14	20	9	11
6	5	30	6	5	12	17	8	9
6	8	48	6	8	10	18	7	11
7	7	49	7	7	8	15	6	9
7	10	70	7	10	6	16	5	11
7	9	63	7	9	4	13	4	9

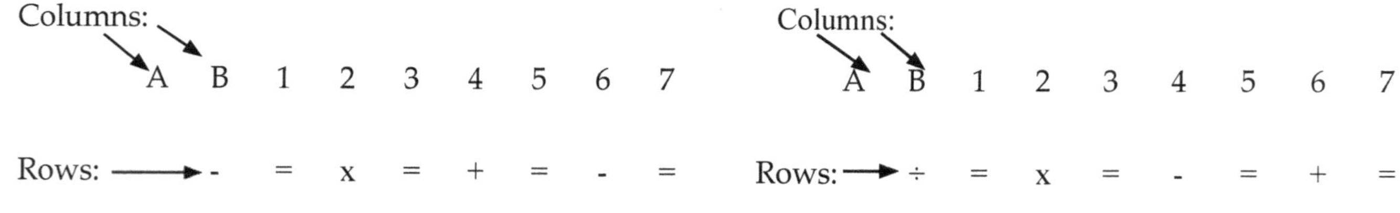

73 — Columns A, B; Rows: - = x = + = - =

A	B	1	2	3	4	5	6	7
20	5	15	2	30	6	36	9	27
18	7	11	5	55	8	63	10	53
16	4	12	4	48	10	58	8	50
14	6	8	7	56	12	68	9	59
12	3	9	6	54	14	68	7	61
10	5	5	9	45	16	61	8	53
8	2	6	8	48	18	66	6	60
6	4	2	11	22	20	42	7	35
4	1	3	10	30	22	52	5	47

75 — Columns A, B; Rows: ÷ = x = - = + =

A	B	1	2	3	4	5	6	7
30	2	15	10	150	10	140	2	142
27	3	9	9	81	9	72	4	76
24	2	12	8	96	8	88	6	94
21	3	7	7	49	7	42	8	50
18	2	9	6	54	6	48	10	58
15	3	5	5	25	5	20	12	32
12	2	6	4	24	4	20	14	34
9	3	3	3	9	3	6	16	22
6	2	3	2	6	2	4	18	22

Walkerdoo Math Puzzles Answer Key- Hard

Note the pattern in Columns A and B first. Then perform the math operation in each row across and fill in each blank cell

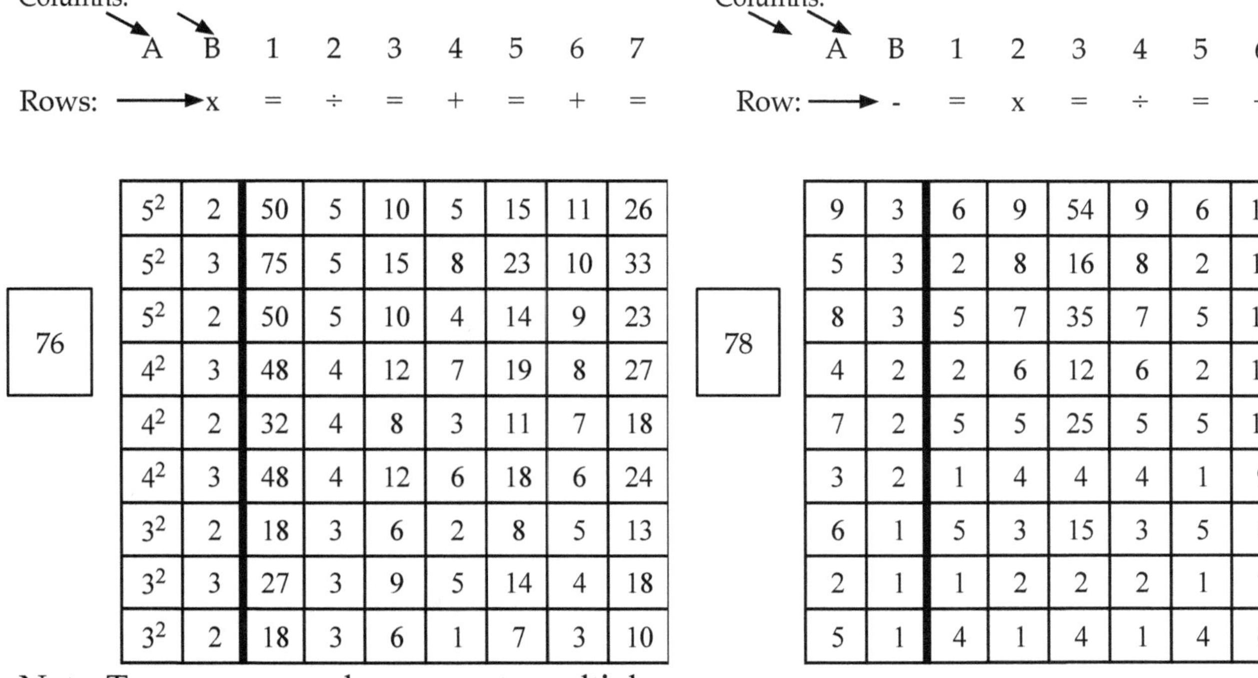

Puzzle 76

Columns: A B 1 2 3 4 5 6 7

Rows: x = ÷ = + = + =

A	B	1	2	3	4	5	6	7
5^2	2	50	5	10	5	15	11	26
5^2	3	75	5	15	8	23	10	33
5^2	2	50	5	10	4	14	9	23
4^2	3	48	4	12	7	19	8	27
4^2	2	32	4	8	3	11	7	18
4^2	3	48	4	12	6	18	6	24
3^2	2	18	3	6	2	8	5	13
3^2	3	27	3	9	5	14	4	18
3^2	2	18	3	6	1	7	3	10

Note: To square a number means to multiply the number by itself. For example, 5^2, is 5 x 5 equals 25.

Puzzle 78

Columns: A B 1 2 3 4 5 6 7

Row: - = x = ÷ = + =

A	B	1	2	3	4	5	6	7
9	3	6	9	54	9	6	14	20
5	3	2	8	16	8	2	13	15
8	3	5	7	35	7	5	12	17
4	2	2	6	12	6	2	11	13
7	2	5	5	25	5	5	10	15
3	2	1	4	4	4	1	9	10
6	1	5	3	15	3	5	8	13
2	1	1	2	2	2	1	7	8
5	1	4	1	4	1	4	6	10

Puzzle 77

Columns: A B 1 2 3 4 5 6 7

Rows: ÷ = - = x = + =

A	B	1	2	3	4	5	6	7
54	2	27	9	18	2	36	4	40
48	3	16	8	8	2	16	5	21
42	2	21	7	14	2	28	6	34
36	3	12	6	6	3	18	7	25
30	2	15	5	10	3	30	8	38
24	3	8	4	4	3	12	9	21
18	2	9	3	6	4	24	10	34
12	3	4	2	2	4	8	11	19
6	2	3	1	2	4	8	12	20

Puzzle 79

Columns: A B 1 2 3 4 5 6 7

Rows: x = + = - = + =

A	B	1	2	3	4	5	6	7
6	7	42	15	57	10	47	3	50
9	5	45	14	59	9	50	4	54
5	6	30	13	43	8	35	5	40
8	4	32	12	44	7	37	6	43
4	5	20	11	31	6	25	7	32
7	3	21	10	31	5	26	8	34
3	4	12	9	21	4	17	9	26
6	2	12	8	20	3	17	10	27
2	3	6	7	13	2	11	11	22

Walkerdoo Math Puzzles Answer Key– Hard

Note the pattern in Columns A and B first. Then perform the math operation in each row across and fill in each blank cell

Columns: A B, Rows: + = - = - = + =

80

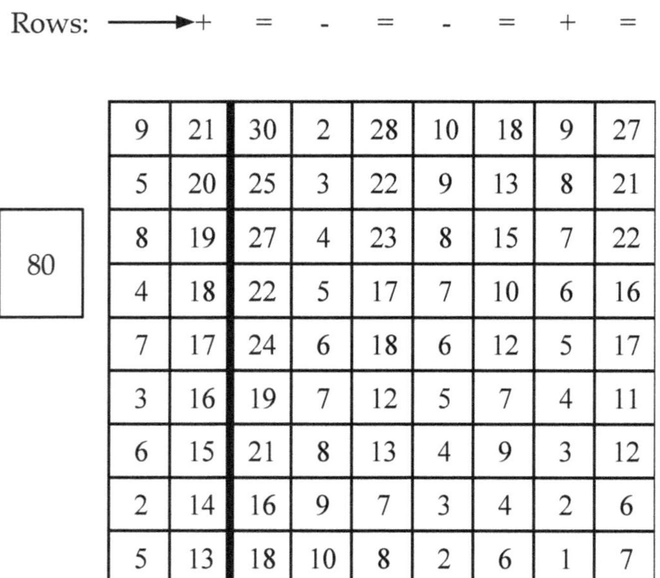

A	B	1	2	3	4	5	6	7
9	21	30	2	28	10	18	9	27
5	20	25	3	22	9	13	8	21
8	19	27	4	23	8	15	7	22
4	18	22	5	17	7	10	6	16
7	17	24	6	18	6	12	5	17
3	16	19	7	12	5	7	4	11
6	15	21	8	13	4	9	3	12
2	14	16	9	7	3	4	2	6
5	13	18	10	8	2	6	1	7

Columns: A B, Row: + = - = x = + =

82

A	B	1	2	3	4	5	6	7
12	13	25	10	15	3	45	2^2	49
8	10	18	9	9	3	27	2^2	31
10	11	21	8	13	3	39	2^2	43
6	8	14	7	7	4	28	3^2	37
8	9	17	6	11	4	44	3^2	53
4	6	10	5	5	4	20	3^2	29
6	7	13	4	9	5	45	4^2	61
2	4	6	3	3	5	15	4^2	31
4	5	9	2	7	5	35	4^2	51

Note: To square a number means to multiply the number by itself. For example, 5^2, is 5 x 5 equals 25.

Columns: A B, Rows: + = x = ÷ = + =

81

A	B	1	2	3	4	5	6	7
15	9	24	1	24	4	6	2^2	10
12	8	20	1	20	4	5	2^2	9
13	7	20	1	20	4	5	3^2	14
10	6	16	2	32	4	8	3^2	17
11	5	16	2	32	4	8	4^2	24
8	4	12	2	24	4	6	4^2	22
9	3	12	3	36	4	9	5^2	34
6	2	8	3	24	4	6	5^2	31
7	1	8	3	24	4	6	5^2	31

Columns: A B, Rows: ÷ = x = + = - =

83

A	B	1	2	3	4	5	6	7
77	7	11	2	22	15	37	7	30
70	7	10	3	30	14	44	5	39
63	7	9	4	36	13	49	6	43
56	7	8	5	40	12	52	4	48
49	7	7	6	42	11	53	5	48
42	7	6	7	42	10	52	3	49
35	7	5	8	40	9	49	4	45
28	7	4	9	36	8	44	2	42
21	7	3	10	30	7	37	3	34

References

Brooker, H., Wesnes, K. A., Ballard, C., Hampshire, A., Aarsland, D., Khan, Z., Stenton, R., Megalogeni, M., & Corbett, A. (2019). The relationship between the frequency of number-puzzle use and baseline cognitive function in a large online sample of adults aged 50 and over. *International Journal of Geriatric Psychiatry, 34*(7), 932-940. https://doi.org/10.1002/gps.5085

Corbett A, Williams G, Creese B, Hampshire A, Palmer A, Brooker H, Ballard C (2024). Impact of Short-Term Computerized Cognitive Training on Cognition in Older Adults with and Without Genetic Risk of Alzheimer's Disease: Outcomes from the START Randomized Controlled Trial. *J Am Med Dir Assoc, 25*(5), pp. 860-865.

The Healthline Editorial Team. (2019, May 22). *Sudoku or crosswords may help keep your brain 10 years younger.* https://www.healthline.com/health-news/can-sudoku-actually-keep-your-mind-sharp